Watergate

A SPECIAL REPORT AT THE REQUEST OF THE

SENATE SELECT COMMITTEE

ON PRESIDENTIAL CAMPAIGN ACTIVITIES

BY A PANEL OF THE

NATIONAL ACADEMY OF PUBLIC ADMINISTRATION

———

Frederick C. Mosher, CHAIRMAN
Alan K. Campbell
Frederic N. Cleaveland
Thomas W. Fletcher
Bernard L. Gladieux
Roger W. Jones
Harvey C. Mansfield
John D. Millett
James M. Mitchell
Harold Seidman
Robert F. Steadman
Richard E. Stewart

Watergate

Implications for Responsible Government

BY

Frederick C. Mosher and Others

WITH A FOREWORD BY

Senator Sam J. Ervin, Jr.

Basic Books, Inc., Publishers

NEW YORK

© 1974 by Basic Books, Inc.
Library of Congress Catalog Card Number: 74–79499
SBN: 465–09085–0 (cloth)
SBN: 465–09702–2 (paperback)
Manufactured in the United States of America
DESIGNED BY VINCENT TORRE
75 76 77   10 9 8 7 6 5 4

CONTENTS

FOREWORD

WATERGATE has become a symbol for corruption in government. It has thus contributed to a growing cynicism in the American people about those in elected or other public office. I consider Watergate to be one of the great tragedies of our history, overshadowing Teapot Dome as well as scandals in the nineteenth century.

That the combination of events known as Watergate, events that far transcend what once was casually labelled a "third-rate burglary," could have occurred should result in the most sober reflection by all thoughtful Americans about the nature of their government. For it is *their* government that was involved; after all, the Constitution of the United States begins with the memorable words, *We, the people of the United States . . . do ordain and establish this Constitution. . . .* And it is the people who have suffered from the Watergate derelictions.

I was honored indeed that my colleagues in the Senate called upon me to chair the Select Committee on Presidential Campaign Activities (the Watergate Committee). It was a great privilege to participate directly in helping to cleanse the Augean Stables of governmental corruption in this country. I am proud of the way in which the other six members of the Committee worked so diligently and cooperated so well during the long months in 1973 when the investigation was beginning and when hearings were held. And I am proud of the dedicated effort by the staff members of the Committee, men and women who worked many long hours under adverse conditions in developing volumes of testimony.

The result produced surely was an invaluable seminar in public enlightenment, not only for the United States but for many parts of the world. The Committee operated in full public gaze, both by the newspapers and by television and radio. One unfortunate consequence may have been a contribution to a growing distrust in government. Surely, however, there was a positive side to the hearings —that of contributing to the store of knowledge people have about their government. I have an abiding faith that with that knowledge will come a fuller understanding of what takes place in Washington —both in good as well as in bad times—that eventually will lead to improvements in the way in which the federal government operates.

The Committee was not alone, of course, in this development of a new awareness in the people. Other governmental institutions have also contributed, not least the courts and the office of the Special Prosecutor. It, furthermore, took a courageous newspaper, manned by dedicated reporters and editors—the *Washington Post*— to spark the entire investigation and subsequent disclosures. Other newspapers and magazines, plus the broadcast media, have contributed greatly to the final result.

In net, then, it is my firm judgment that Watergate helps to prove the essential soundness of our institutions and of our Constitution. Once the people were made aware of what had taken place, and what at this writing (April 1974) continues to occur as new improprieties are disclosed, their response has been heartening to me. I believe that I can detect a growing insistence throughout the country that there be basic improvements made without delay. The thousands of letters and telegrams received by the Committee are evidence of that. The tide of reform is running strong and deep.

If it did nothing else than enlighten the public, the Watergate Committee's work can only be considered to be a great success. One of the primary functions of a Congressional committee is to inform the electorate, a principle that has become embodied in our constitutional law. That the electorate was informed, and through that process of information there were set in motion strong and continuing movements for fundamental change, cannot realistically be denied.

One of the most important of those changes is the restoration and maintenance of balance in our Constitutional system—in what is called the separation of powers. No one, certainly not I, wants an impotent Chief Executive. We have, however, witnessed in recent years—not all in the present Administration—a dangerous imbalance of power tilting away from the Congress toward the President. Congress no doubt has contributed to that condition. Watergate gives the Congress—and the people as a whole—an unparalleled opportunity to redress that power tilt toward the White House. A similar opportunity will not soon come again.

In my judgment, it is imperative that steps be taken without delay, but only after due study and consideration, to achieve the twin goals of placing adequate checks upon any Chief Executive while simultaneously permitting him to provide the leadership for which he was elected. If we are to move toward realization of the rule of law in this country, an ideal that has become so badly tarnished in recent times, then this task must be undertaken and achieved. The Constitution did not establish Presidential government, unchecked by law or by the other branches, but we have slipped dangerously toward precisely that. Rather, the Constitution set up a system by which separate institutions shared governing power, and thus were able to prevent the undue concentration of power in any one branch.

In its work of investigation and the development of what took place during the Presidential campaign of 1972, the Committee and its staff have always kept in mind that their second primary function was to produce recommendations for legislative change in our laws. These recommendations are being set forth in detail in the final report of the Committee. With the filing of its report, the Committee's work will cease—but others must then undertake study and considerations of those recommendations.

Our efforts in writing the report have been helped immeasurably by a number of scholars and organizations, all of whom gave freely of their time to send ideas for consideration by the Committee. Not least among these sources was the National Academy of Public Administration, which, at the request of the Committee undertook, under forced draft, to render a report to the Committee on the implications of Watergate for our democratic system of government. We have benefited a great deal from this effort by the Academy, an effort, I might add, that was accomplished without cost to the taxpayer. I am pleased that the Academy's product is now being published, for it deserves a wide readership. In it may be found some of the most thoughtful commentary about the government of the United States that I have ever read. The Academy, particularly the special panel convened to make the study, has my deep appreciation for its unselfish, patriotic endeavor.

Sam J. Ervin, Jr.
United States Senator

April 1974

PREFACE

ON October 5, 1973, Senator Sam J. Ervin, Jr., Chairman of the Senate Select Committee on Presidential Campaign Activities, wrote the National Academy of Public Administration to invite its views about the basic issues revealed in, and arising from, the Watergate affair for the Committee's consideration in preparing its own report. More specifically, the Senator asked the Academy to help the Committee to ". . . identify the major implications developed by the hearings." He stated, "It would be extremely helpful to us if you could let us have the judgment of your organization on the following matters: (1) what are the key issues or problems in our governmental system that have been disclosed by the Committee's work? and (2) what are the options or alternatives that might feasibly be open for serious contemplation by the Committee and its staff in writing its recommendations? With respect to the latter question, what are the advantages and disadvantages of each alternative?"

The Academy, through its Board of Trustees, agreed that such an undertaking was fully in keeping with its basic mission, and designated a 12-member Panel to study, and prepare a report on, the governmental implications of Watergate. The Panel was chosen to provide a broad spectrum of experience in administration and of scholarship in the field. Most of its members have held high administrative positions in government, principally, but not exclusively, at the national level, and several have held administrative posts in nongovernmental kinds of enterprises—business, education, and others. The majority also have taught and written in public administration. The Panel is a nonpartisan, or multipartisan, group. None of its members is active in partisan politics, which was no consideration in the deliberations of the Panel.

The report which follows is essentially a Panel report, not one prepared for a committee by its staff. The Panel met as a whole three times over a total period of about five days. There were countless smaller meetings of members and outsiders on specific problems. Every member was assigned a particular task of writing a basic paper in a field in which he had particular expertise, and all of these documents, following discussion and revision, provided the basis for the report. The document itself is a product of the nearly full-time efforts to integrate, coordinate, and edit these materials by the Panel's Chairman, Professor Mosher, the Staff Director, Bertrand Harding, and Richard Chapman and Erasmus Kloman of the Academy staff. The report includes a minimum of footnotes and references to other works, since it is designed as an action, and not an academic, document. All of the members of the Panel were consulted before the report was issued. Despite questions and criticisms on various particulars in language, the report is approved unanimously by members of the Panel, with no dissents. In the interest of clarification and editorial improvement, the original version of this document has been somewhat modified in form, but not in substance, for purposes of this publication.

The Panel received advice and assistance from a great many other persons, both inside and outside of the Academy, and it wishes to express its appreciation to all of those who contributed. Mitchell Rogovin, of Arnold and Porter, and Robert Layton, of Layton and Sherman, provided advice and counsel on legal aspects of the study. A five-member committee—Morris W. H. Collins, Jr., Dean of the College of Public Affairs, The American University; James W. Fesler, Cowles Professor of Government, Yale University; Harold B. Finger, Manager of the Center for Energy Systems, General Electric Company; William T. Golden, Corporate Director and Trustee, New York City; and Ronald B. Lee, Director of Marketing Analysis, Xerox Corporation—reviewed and criticized the manuscript in its final stages. Miss Barbara Byers edited the report and Mrs. Carolyn Holliday typed the innumerable drafts and the final manuscript.

Preface

The Academy is especially grateful to The Ford Foundation and to John D. Rockefeller III for the financial support which made this study possible.

In publishing a study the Academy presents it as a competent treatment of a subject worthy of public consideration. The interpretations and conclusions in such publications are those of the members of the cognizant Panel and do not necessarily reflect the views of the officers or other members of the Academy.

Roy W. Crawley
Executive Director

March 1974
National Academy of Public Administration
1225 Connecticut Avenue, N.W.
Washington, D.C. 20036

Watergate

INTRODUCTION

Overview

PRACTITIONERS and scholars in the field of public administration have an extraordinary interest in the quality of governmental institutions. They share with all citizens a concern that the competence and dependability of governments be achieved and secured. But, beyond this, those who have committed most of their lives to public service—advisers on public policy, practitioners sworn to faithful execution of the laws, and educators of present and future administrators—feel a special obligation to preserve the values that have so long contributed to an effective and progressive social order. They particularly appreciate the absolute necessity of integrity of the leaders in every branch of government—legislative, executive, and judicial. Without such integrity, government cannot gain and retain the confidence of the people it serves.

For these reasons, this Panel of the National Academy of Public Administration is gratified and challenged by the invitation of the leaders of the Senate Select Committee on Presidential Campaign Activities to present its views on issues emerging from the Committee's hearings. The revelations, immediately or remotely associated under the umbrella term "Watergate," have had a shattering impact upon American government at all levels. They have played a major role in causing the citizenry to develop, and to give voice to, growing disillusionment, cynicism, and even contempt for government and politics generally. But there is also a potentially favorable side. The very dimensions of the scandals so far revealed provide an opportunity for reexamination and reform, not of the electoral process alone, but also of other related practices and institutions.

Some of the seeds which grew to Watergate were undoubtedly planted many years and many administrations ago. But the development of these seeds into malpractices seriously injurious to our democratic form of government calls for sober reevaluation of our political and administrative systems and the application of appropriate remedies.

This report is not directed to the identification of individual misdeeds or culprits. Rather it is an effort to identify underlying sources and pitfalls, and to suggest changes in American government and administration which will help make future Watergates less likely, and which will improve the effectiveness and credibility of democratic government over the long range.

Aberration, Extension, or Culmination?

Are the various deviations from proper behavior that are popularly associated with Watergate to be regarded as one-time events, the product of a particular combination of circumstances and of people, mostly at high levels, in a political organization and in the Administration? Were they unique in American history and unlikely to recur in the future?

Or was Watergate simply an extension of trends in American politics and government that have been underway for a long time and which could, unless deliberately checked or reversed, be expected to continue, and even worsen, in the future? More profoundly, is it a reflection of developments and deterioration in the very fabric of American society, social, economic, moral, and technological?

Or, finally, was Watergate a cataclysmic shock, a peaking of the trends and forces suggested above, from which society and government may not recover without severe surgery?

It appears to this Panel that Watergate permits all three of these interpretations. Surely it was an aberration in the sense that it resulted from bringing into positions of enormous power a group of people who shared characteristics of personal and ideological loyalty and inexperience in social responsibilities. Surely it will not soon recur—not necessarily because such a collection of people may not again be assembled, but because of the disastrous consequences for many of those involved individually and for the Administration generally. Surely it is not the first scandal which sullied American public life. We have had Credit Mobilier, Teapot Dome, political corruption of public relief programs in the 1930s, and Internal Revenue malfeasance in the early 1950s, although none of these had the pervasiveness and shattering impact of Watergate.

On the other hand, one may observe that most of the perpetrators and directors of Watergate misdeeds were reputedly honest and upright persons before they entered the political campaign and/or the Administration. Few if any had any record of unethical or dishonest, let alone criminal, activity. This suggests that Watergate was a product of a *system* which shaped and guided the behavior of its participants. (System, as the word is used here, is primarily the product of trends and forces from the past.)

President Nixon, as well as others in his Administration, has defended some of his and their actions on the grounds that the same things were done by predecessors in high office. Regardless of the validity of this contention as defensive argument, it is entirely true that many of the actions deplored in this report and elsewhere have precedents in previous administrations of both parties. Indeed, the evidence suggests that most of such practices were growing gradually or sporadically during past decades. They include, for example: (1) use of governmental powers and resources in behalf of friends, against opponents (enemies); (2) politicization of the career services; (3) political espionage on American citizens; (4) excessive secrecy, usually on grounds of national security, whether or not justifiable; (5) use of governmental personnel and resources for partisan purposes, including political campaigns; (6) solicitation of political contributions from private interests with implicit or explicit assurances of support or favor, or of absence of disfavor; and (7) "dirty tricks."

Some have argued that Watergate was a logical, if not inevitable, consequence of trends in the larger systems of our society.

Thus one reads and hears that its forebears include, among many other things: the weakening of the family unit and with it the sense of responsibility to and for others; the decline of personal and, therefore, of social morality and of the influence of the churches; the growing interdependence of major elements of the society and of the economy, and of the government with both of these; the increasing power of the national government; the growing interrelationships of the national government with other governments in the world; the increasing prevalence in both public and private spheres of huge bureaucracies, in which the individual is submerged; and the growing dominance of technology.

It would be both inappropriate and impossible for this Panel to address these alleged negative trends in the space and time permitted. The Panel does, however, agree that forces such as these may have contributed to Watergate. But it also believes that government, particularly the national government, has been in the past, and should be in the future, strong, beneficent, and flexible enough to influence these forces toward the benefit of the American people. The ethical and effective conduct of government must provide the model and the leadership for American society.

Watergate is thus both an aberration and an extension of earlier trends. It may also be a culmination of some, if not all, of those trends. As suggested earlier, the revelations in the 12 months of 1973 should themselves deter possible future perpetrators from at least the most blatant of such misdeeds. Certainly, these revelations have alerted the American people and their elected representatives to the danger of future Watergates and initiated a search for legal and other means to thwart or minimize them. Many basic reforms of American government, including the framing of the Constitution itself, have been sparked by conspicuous failings, scandals, tragedies, or disasters.

The Watergate revelations have already stimulated a great deal of concern and discourse about means of political and governmental reform. And there is reason to believe that the Congress and the people whom it represents are more receptive to basic changes than they have been in a long time. The "horrors" which have been, and are still being, exposed have a potentially posi-

6

tive side. They offer an opportunity for corrective actions, many of which should have been taken long ago. If the opportunity is grasped, Watergate will be more than an extension of long-term trends; it will be truly a culmination and watershed.

The Watergate Climate

Some of the witnesses before the Select Committee spoke about the unique and, it would appear, altogether unpleasant "climate" which pervaded the top levels of the Administration and the Committee to Reelect the President during and before the Watergate period. In the words of John Dean, Watergate was an "inevitable outgrowth of a climate of excessive concern over the political impact of demonstrations, excessive concern over leaks, and insatiable appetite for political intelligence all coupled with a do-it-yourself White House staff regardless of the law."

The Watergate climate, for convenience, may be treated from two perspectives, the political and the administrative, even though the close interrelationship of the two was one of its central and most sinister features. The prime motivating drive behind both political and administrative activities seems to have been Presidential *power,* its enlargement, its exploitation, and its continuation. Power was perhaps sought by some in the Presidential entourage for its own sake, but it seems fair to conclude that most sought to impose upon the government the ideological views of the President. Paradoxically, a part of that ideology as expounded was to limit the powers of the national government; to return more powers to the people and their elected representatives at state and local levels. The zest for power in the Presidential office is a perfectly expectable, normal, and proper behavior of Presidencies, particularly in the current century—but within limits and constraints, many of which are embedded in the Constitution, and always subject to accountability.

The political climate in the months and, to a lesser extent, for several years before the 1972 election was apparently characterized by an obsessive drive for reelection of the President. It seems to have colored, or sought to color, governmental plans, decisions, and behavior during that period, even in fields of activity intended and believed to be politically neutral—administration of the revenue laws, antitrust prosecutions, allocation of grants and contracts, clearance of career service appointments and promotions, and many others.

Synchronized by the White House and its immediate appendage, the Committee to Reelect the President, the program was directed primarily, if not almost exclusively, to the reelection of the President, not to the victory of his party or of other nominees of the same party in the same general election. It is evident that the imperative to reelect was so driving as to override many other considerations, including the public interest and normal ethical and legal constraints.

The President, in explaining the behavior of his subordinates, described them as "people whose zeal exceed their judgment, and who may have done wrong in a cause they deeply believed in to be right." But some of those whose behavior the President described so mildly readily admitted later that, in their minds and consciences, the demand to reelect was so overwhelming as to justify acts admittedly criminal. John Mitchell, in his testimony before the Select Committee, made it clear that he considered the reelection of President Nixon to be more important than his obligation to tell the President that people around him were involved in perjury and other crimes, even though he could find no Constitutional basis for such a conclusion.

The political environment, both before and following the campaign of 1972, was entirely consonant with that of the campaign itself. The Administration was in a state of siege from its critics, some of whom were seen as threats not only to it alone, but also to the security of the nation itself. Administration proposals were advanced, the potential costs of which so far exceeded their potential benefits, even in strictly political terms, as later to be construed as stupid, even absurd. The White House became a command post for conduct by the President's staff of near warfare against those whom it considered "enemies."

Following the "mandate" of the 1972 election, the Administration moved to cleanse itself of senior officials in many executive agencies who were considered to be hesitant or doubtful followers of the views and ideology of the President. In terms of top-level political appointees, the transition between the first and second Nixon terms was as extreme as most transitions from one party to the other. Many experienced Republicans in key posts were replaced by others, usually younger, in whom the Administration presumably had greater confidence of personal and ideological loyalty, and who were innocent of prior allegiances to the agency of their appointment or its associated clienteles. In this and other ways the Administration undertook to carry out and enforce its electoral "mandate," even before the inauguration in 1973.

The administrative climate was, to some extent, a product of the political climate: aggressive efforts were made to use administrative machinery to carry out political and policy ends, and growing frustration and exasperation developed over alleged bureaucratic impediments. In part, it was a further step in the evolution of a strong Presidency—a movement which had begun generations earlier and which students of American government have generally approved, at least since the report of the President's Committee on Administrative Management (the Brownlow Report) submitted in 1937.

Taken individually, the majority of changes that the Administration instituted or sought were consistent with sound administrative practices; indeed, a good many leaders in public administration had recommended some of them earlier and specifically endorsed them after they were proposed by the President. They included:

- formation of regions with common headquarters and boundaries to encompass many of the domestic field agencies and activities;
- establishment of regional councils to provide better coordination of federal activities in regions and areas;
- delegation of federal powers from Washington to the field;
- unconditional grants to state and local governments (called general revenue sharing);
- broader categories of, and fewer strings on, functional grants to state and local governments (called special revenue sharing);

- formation of a Domestic Council to parallel, in domestic affairs, the National Security Council in foreign affairs;
- strengthening of the managerial role of the Bureau of the Budget (which became the Office of Management and Budget), and vesting of all of its statutory powers in the President;
- consolidation of the activities of most of the domestic departments in four "superdepartments," rationally organized according to subject matter areas;
- formation of a "Federal Executive Service" to encompass all supergrade employees whose qualifications would be approved and whose assignments and salaries would be flexible according to managerial needs;
- encouraging the administrative practices associated with the term "management by objective";
- placing postal activities in a quasi-governmental corporation and removing the Postmaster General from the Cabinet; and
- interposing above the specialized, professionalized, "parochial" bureaucracies generalists with a broader perspective.

But these mostly constructive actions and proposals were accompanied by a number of others which students of government, even those with the strongest commitment to Presidential energy and influence, found questionable. These included:

- usurpation by the White House of powers over both policy and day-to-day operations heretofore carried on in the departments and other established agencies;
- enormous growth of the White House staff, accompanied by the establishment of a tight hierarchy within it;
- by-passing of departments and agencies in areas of their assigned responsibilities, first in international and defense matters through the staff director of the National Security Council, and later through the staff director of the Domestic Council;
- veiling of White House activities on grounds of national security or executive privilege;
- negating of substantial majorities of both houses of the Congress on policy and program matters through accelerating use of the veto power and impoundment of funds;
- interposition of White House aides between the President and the official heads of the executive agencies, such aides having been appointed without confirmation or even public knowledge; and
- the abortive attempt to interpose Presidential counselors in the White House with substantial control over established departments.

Considered singly or separately, few of the actions or proposals in the foregoing lists would be cause for great alarm. However, if all of them had been effectuated, the administrative weather could have become very stormy indeed. The American state then would have approached a monocracy, ruled from the top through a strictly disciplined hierarchical system. It would have become difficult to pin responsibility for decisions or actions upon anyone short of the top man, and he was, for the most part, inaccessible and unaccountable. As some of his appointees have pointed out, the only ultimate means of holding the President answerable following his election or reelection is impeachment.

The administrative and political aspects of the governmental climate were increasingly interlocked, at least until the spring of 1973. Together they constituted a critical threat to many of the values and protections Americans associate with a democratic system of government, including:

- the right to participate or be represented in decisions affecting citizens;
- the right to equal treatment;
- the right to know;
- free and honest elections;
- assurance of Constitutional protections such as those in the First and Fourth Amendments;
- a balance of countervailing powers to prevent usurpation by any single power—as among the branches of government, the political parties or sectors of parties, interest groups, and geographic sections; and
- ethical conduct of public officials in pursuit of the public interest.

This Panel entertains no delusions that these precepts have not been violated or threatened before—well before—the first Nixon Administration. It is possible that many of the revelations of the last two years were repetitions of earlier actions never revealed. Indeed, we owe our present knowledge to an obscure and lonely guard in the Watergate Apartments, to the fact that there was an unmarked police car nearby, to a couple of aggressive reporters, to the Senate Select Committee, to the Special Prosecutors, and to some unidentified leaks to the press. The problems we here address are not partisan problems. They are problems for all parties, for all citizens.

Public Knowledge and Understanding

As Thomas Jefferson and a good many others have emphasized, an effective democracy depends upon an alert, educated, and knowledgeable citizenry. The operation·of a responsive and participatory system of government in a large, variegated, interdependent, and complex society is one of the most difficult and sophisticated problems ever faced by mankind. It requires a body of officials sensitive and responsive to the wishes and needs of citizens and dedicated to the service of those citizens. And it demands channels of information and access between government and governed in order that the public service may respond to those needs.

In the opinion of this Panel, altogether too little, and declining, attention has been paid to the ideals and the workings of democratic government in the educational system—the schools, colleges, universities, and professional schools—and in the media. Recent assessments of knowledge about government among high school students are very disappointing. Higher education has increasingly dedicated its efforts to science and methodology and away from the normative considerations essential to democratic choice making. Professional education, which now comprises a very large part of higher education, is almost devoid of content on the special problems of public service in a democratic state, even though more than one-third of professional graduates are destined for careers in politics and government. And the media, with their concentration on crime, scandal, commercial advertising, sports, and other entertainment, have fallen far short of their educational potential.

One of the most important contributions of the Senate Select Committee's televised hearings was its education of millions of American citizens about the workings of the national government. Unfortunately, they were taught a great many things that should not have been there to learn. And this too is a reflection on our educational system. Very few of the top witnesses indicated any sense of understanding or appreciation of democratic ideals or principles. Almost none mentioned any special considerations of *public* service for the *public* interest apart from the President's interest. They had not learned in secondary school or in college or in law school that there is something special and different about public office and public responsibility.

Recent polls indicate the appalling lack of knowledge of an enormous number of Americans about their government. Yet all depend upon it in a number of ways; as recent as well as earlier history has abundantly demonstrated, government is vitally important to all of them. In any nation which aspires to democratic principles, government is the people's business. The Watergate scandals must have whetted the curiosity of a good many about government, at least about what went wrong. The greatest value of the Senate Select Committee's final report may be its contribution to education of the public about American government. And the Academy's Panel, in submitting this report, dares to hope that it too may have some educational value.

This Report: Its Objectives, Premises, and Content

In the pages which follow, the Panel has undertaken to raise and address what appear to it to be the key questions related to or behind the Watergate affair; to consider alternative options, where alternatives seem worthy of consideration; and, where there was substantial consensus among the Panel members, to make recommendations accordingly.[1] Some of the considerations and discussion in the report go well beyond the questions and testimony of the Senate Select Committee's hearings because the implications for government and administration are far broader and deeper than the statements of individual witnesses. On the other hand, the limited time available has not permitted any extensive re-

[1] For purposes of this report, "consensus" is considered to have been reached when no more than two of the twelve panel members disagreed.

search, so the Panel has tried to focus its attention only on what
it considers most relevant in terms of the Watergate hearings.

The entire report is premised upon a number of basic convictions which were discussed, considered, and unanimously agreed
upon early in the Panel's deliberations. They are:

1. A public office is a public trust.
2. All public officials, like other citizens, are subject to the laws of the
   nation.
3. The proper conduct of the nation's affairs necessitates strong executive leadership.
4. The proper conduct of the nation's affairs necessitates a strong
   Legislative Branch, with effective oversight capability.
5. The proper conduct of the nation's affairs necessitates acceptance
   of the finality of the role of the judiciary in interpretation and
   application of the laws.
6. The proper conduct of the public's business requires maximum
   openness, consistent with the nation's security and individual
   privacy.
7. The highest ethical standards are an essential component in the
   conduct of public business.
8. There must be a responsive public service, with appropriate balance between a high quality political staff and strong, professional
   career services.
9. Public officials should be continuously accountable to the public
   for their official acts, and there must be effective mechanisms for
   maintaining that accountability.
10. The primary loyalty of all public servants must be to the American
    people, through the institutions of government, and not to any in-
    dividual leader.
11. An open, free, and honest electoral system is fundamental to a
    democratic society.
12. Effective government requires a strong, responsible, and competi-
    tive party system.

The Panel recognizes that these premises are guides and criteria,
dependent ultimately upon the conduct, judgment, conscience, and
relationships of people. There are limits to what can be accomplished through an Executive Order or Reorganization Plan, a
statute, even a Constitutional amendment. One cannot legislate
ethical behavior, although laws may and do set outer boundaries

of permissible behavior and provide incentives for compliance—as well as penalties for violation. Many of the most useful, even essential, expressions in the governmental lexicon are undefinable in advance of specific cases (and not only legal cases): due process, public interest, national security, responsiveness, and coordination. Some of the options considered in this report involve statutory change; a few, Constitutional amendment. But many are exhortations that those in public office or seeking public office may behave responsibly and with devotion and integrity.

CHAPTER 1

The American System
of Government

THE WATERGATE revelations have already provoked a rash
of proposals by scholars, politicians, journalists, and others to
change the United States governmental system. And there will
surely be more to come. Some of these would go to the roots of
the Constitution and the political philosophy on which it is
grounded. They rest on a belief that Watergate was the product
of a fatal flaw in the system, or that the overall design of government wrought for the eighteenth century is simply not adequate
to meet the demands of the late twentieth century, or both. Less
radical proposals would correct or modify certain more or less
specific practices or institutions by Constitutional amendment or
by statute. This Panel has considered and discussed the more
prominent of these proposals.

Toward a Parliamentary System?

The majority of the fundamental changes which have been advanced would move American government toward, or replace it
with, a parliamentary system, and thus significantly modify the
existing Presidential-Congressional division and sharing of powers.
Some might go all the way to a British type of government in

which the executive is the leader of the majority party in the Parliament and subject to removal by Parliament. Most of the proposals that have come to this Panel's attention are partway houses: provisions for dissolution of the government by Congress or by President and Congress; provision for a vote of no confidence to remove the President and dissolve the Congress; a division of the President's powers between a chief of state and a chief of government (presumably a prime minister); a council of state, including representatives of the Congress; a collegial executive; an automatically and frequently rotating executive; and a constitutionally required question period before Congress for the President and his Cabinet.

The arguments behind these proposals are varied. Central to most of them is that the Presidency has become too strong, the Congress too weak. Second is that it is too difficult to hold a President accountable for his acts. Another is that the calendar system of elections is too rigid in a rapidly moving society to provide for political responsiveness. Finally, some feel that the Congress is closer to its many constituencies, more responsive, and more accountable, and therefore should have the ultimate power.

While acknowledging validity in some of these positions, this Panel is convinced that the nation needs both a strong, single Chief Executive and a strong Congress. The weaknesses of the Congress derive not from the Constitution but from problems of its own organization and practices, most of which are correctable. Despite one devastating Civil War, the Constitution and the premises on which it rests have on the whole served the nation well for 185 years. It is broad enough, flexible enough, and amendable enough to permit growth and adaptability to rapidly changing demands in a dynamic society. It has survived the severest of strains, as it is surviving those of today, and, it would appear, has usually been strengthened by them.

As political philosophers back at least to Socrates have noted, no system of government can provide absolute assurance against corruption or incompetence in government. Watergates could have occurred and could have been covered up under a parliamentary system. Had it not been for the independence of the Congress, through the Senate's Select Committee and the House Judiciary Committee, the independence of the courts, and the freedom of the "fourth branch," the press, Americans might never have learned of Watergate. The uncovering has been a long, slow, divisive process, and it is still unfinished. But without the separation of powers, it might never have occurred at all.

The Panel notes too that, with the growing complexity and growing importance of national governments around the world, the need for a unified executive, adequately staffed, has led parliamentary governments, especially in Europe, to move in the direction of Presidential systems, not the reverse.

The Panel is unanimously opposed to basic changes in the American Constitution in the direction of parliamentary government.

Presidential Removal

The only means of separating a President or Vice President from his office during his elected term (aside from death) are voluntary resignation, impeachment, or, under the twenty-fifth Amendment, inability "to discharge the powers and duties of his office." In the last case, the inability may be declared either by the President himself or by a group of high officials; in either instance the office is assumed by the Vice President as Acting President until such time as the President is declared, or declares himself, capable of reassuming his responsibilities.

In the 185 years of Constitutional history, no President has been separated from his office during his term except by reason of death. Except for two resignations, the same is true of Vice Presidents. No elected Federal official has ever been convicted on impeachment. The one effort to impeach a President over a century ago, which failed of conviction by a single vote, grew out of, and added fuel to, a bitter and divisive battle between the President and the majorities in the Congress, between South and North. There are two or three occasions in American history when the "inability" provisions of the twenty-fifth Amendment might have

been invoked, had they been in the Constitution from the beginning, but whether they would have been utilized is unknown.

Obviously, the removal of a President or a Vice President during his term of office against his will has proved difficult, almost impossible.

IMPEACHMENT

The Constitution provides that public officers "shall be removed from office on impeachment for, and conviction of, treason, bribery, or other high crimes and misdemeanors" (Art. II, Sect. 4). It elsewhere defines treason (Art. III, Sect. 3), and bribery has a fairly well-understood meaning. But nowhere does it explicate what constitute "high crimes and misdemeanors." There was, at the time of the Constitutional Convention, virtually no American national criminal law or other precedent, and it is clear that the founders were relying upon English practice and experience, basically in fear of an "elected monarch" in the Presidency. It is therefore essential to look to history, even medieval history, to understand the term's origins and intended meaning.

The original proposal on impeachment during our Constitutional Convention provided as grounds "malpractice or neglect of duty." This was changed by the Committee on Detail to "treason, bribery or corruption" and later to "treason or bribery." George Mason recommended on the floor that "maladministration" be added, but, on James Madison's objections that this was too vague, changed it to the old English term, "high crimes and misdemeanors."

The expression has had a long, and long-abandoned, history in England. "High crimes and misdemeanors" were not necessarily criminal in the ordinary legal sense. In fact, it appears likely that many of the British—and later, American—cases were not grounded on indictable crimes. They derived not from criminal common or statutory law but from Parliamentary "common" law. "What lends a 'peculiar' quality to these crimes is the fact that they are not encompassed by criminal statutes or, for that matter, by the common law cases. . . ." [1]

[1] U.S. Congress, House Committee on Judiciary, *Impeachment: Selected Materials*, H. Doc. 93–7, 93rd Congress, 1st Session, p. 623.

Raoul Berger, perhaps the most eminent American scholar on the subject, summarizes his conclusion from various British cases of impeachment that "high crimes and misdemeanors" included *misapplication of funds, abuse of official power, neglect of duty, encroachment on or contempts of Parliament's prerogatives, corruption, and betrayal of trust.*[2]

The United States has had relatively few precedents of impeachment. (Only 12 reached the Senate; only four, all judges, were convicted.) Several, if not most, did *not* rely upon indictable, criminal offenses, although some were defended on the basis that such offenses had not been committed. Impeachment in Britain and in the United States has always been an essentially political proceeding—and this may explain why it was vested in the Commons and Lords in Britain, and in the House and Senate here. In Britain, conviction on impeachment could involve punishment up to death; the United States provided no punishment other than removal from office, with the understanding that any crimes committed could be further pursued in the courts. This was the one basic change.

Our current problem is one of understanding what "high crimes and misdemeanors," an antiquated expression, means or meant, particularly to the framers of the Constitution. It appears that, in the framers' minds and in subsequent experience, a President or other official may properly be impeached and removed from office for major offenses to the society and the state without being beheaded, jailed, fined, indicted, or even indictable.

This view is supported by the deliberate separation of the processes of impeachment and subsequent punishment. The clumsy process of impeachment fell into disuse when the British, in the eighteenth century, devised a handier way of removing ministers of the Crown on purely political grounds of want of confidence in Parliament. But because the framers of our Constitution rejected legislative dominance in favor of an independent executive and a separation of powers, they retained impeachment and the ambiguities of its historic grounds as an ultimate defense against executive and judicial abuse. They made the

[2] Ibid., pp. 628–629.

process difficult and arduous to prevent its intemperate or trivial use and to assure the Presidency of strength and stability. *This Panel believes this to have been wise and makes no proposal to change the process. But, at the same time, the Panel urges that the American people and their elected representatives be educated to the original intent and meaning of "high crimes and misdemeanors."*

This might conceivably be done by an educational campaign beginning in the Congress itself. It might be done more effectively through a statute defining the meaning of "high crimes and misdemeanors" in modern language such as "major crimes, misconduct in office, or neglect of duty." Or, in the present climate, a Constitutional amendment substituting such language might be both more feasible and more effective. But the Panel urges upon the Select Committee serious consideration of these alternatives.

Some persons have noted that a horse and jockey, if they win, place, or show, are immediately disqualified if they are judged to have violated the rules. The same is true in many other sports. But in the most important competition in American life, a Presidential campaign, a successful candidate, or his campaign organization, can perform almost any illegal action to win his nomination and election without challenge to his right to assume office. It was clear months before the 1972 election that the campaign was attended by illegal activity; only a few months later, the dimensions of such activity began to unfold. Yet there was no way to call into question the validity of the election itself.

The conduct of Presidential campaigns is treated in the closing chapter of this report. But it seems appropriate to consider here the propriety of assumption to office by one whose nomination and election have been won through illegal means. It appears impracticable to pronounce an election void for reasons of both procedure and time.

But it is the consensus of this Panel that the grounds for impeachment be extended to include serious misconduct in the political campaign conducted prior to assuming office. This may require a Constitutional amendment.

RECALL ELECTIONS

It has been suggested that the Constitution be amended to provide for recall elections of a President or Vice President, as in some state and local jurisdictions, by petition of some proportion of voters, or by a majority or two-thirds vote of both houses of Congress, or both. The arguments advanced for such a provision are that it would provide a political alternative for removal, not subject to a quasi-criminal proceeding, and that it would restore the decision as to removal to the same voting forum as the one which originally elected. Against it are the fear that a nationwide recall election would be a traumatic and divisive event, and the hope that simpler means for removal could be found through existing institutions and processes.

It is the consensus of the Panel that provision for recall elections would be undesirable.

The Presidential Term

Two quite opposing proposals have been made to change the provisions as to the Presidential term: one would extend the term to six years and make it nonrenewable; the second would repeal the twenty-second Amendment, which limits a President to two four-year terms (or one and a half in the case of Vice Presidential succession).

In favor of the nonrenewable six-year term, it is argued that:

- it would encourage Presidents, once elected, to view their roles as Presidents for the whole people without concern about reelection, and to behave as statesmen, not politicians;
- it would reduce or eliminate the inherent advantage of the incumbent over his less prestigious and less well-known rivals;
- it would discourage or eliminate the temptation to utilize governmental resources and powers for partisan purposes; that is, it would encourage more impartiality in the conduct of the government's business; and
- it would reduce the frequency of Presidential political campaigns, viewed by some as traumatic experiences, by one-third.

The main argument of opponents of the nonrenewable six-year term is that it would further reduce the sense of public accountability of the Presidency. Particularly in view of the difficulties of removal (discussed in the preceding section), it might even encourage a corrupt, incompetent, or negligent incumbent to misdeeds and prolong the period during which they could be performed. Proponents of this view point out that many of the deplorable actions of the present Administration were performed after the President's reelection was assured and, under the twenty-second Amendment, when he could look forward to no possible reelection.

It is the consensus of this Panel that adoption of a nonrenewable six-year term would be undesirable.

Arguments for and against repeal of the twenty-second Amendment are for the most part the same as the above, although on opposite sides. Proponents of repeal hold that the Amendment itself was a mistake and a somewhat vindictive reaction to President Roosevelt's third and fourth terms. They feel that the possibility of more than two terms would encourage more responsible behavior of a President in his second and subsequent terms. Opponents fear that, with the inherent advantages of incumbency, a President might conceivably build an election machine which would assure almost unlimited Presidencies, not unlike a monarch.

A majority—but not a consensus—of the Panel is opposed to repeal of the twenty-second Amendment, but believes this question deserves the thoughtful consideration of the Select Committee.

Special Presidential Elections

It has been proposed that, in the event of vacancies of both President and Vice President, there be provided a special election for both offices, and a bill to this effect has been introduced in the Congress.[3] In fact, there is some indication that the framers

[3] Remarks by Rep. Joe Moakley on introducing H.R.11214. *Congressional Record*, daily ed., 119, no. 165 (October 31, 1973): E6945.

of the Constitution contemplated it, and the Second Congress enacted a statute providing special elections which survived for nearly a century—until 1886. The constitutionality of such a bill has been attested by a number of qualified lawyers.

The twenty-fifth Amendment provided for Presidential appointment, with Congressional approval, of a Vice President in the event of vacancy in this office. It is unlikely that there would be simultaneous vacancies in both offices. If both President and Vice President died or were removed nearly simultaneously, or if both were removed because of illegal election activities, as suggested above, such a contingency requiring special Congressional action might occur. In such cases, a special election might be necessary.

Opponents to special Presidential elections argue that such elections would disturb the stability of the Presidency and upset the accustomed and calendered regularity and continuity of that office.

The Panel has reached no consensus on special elections. But if its recommendation about termination of office because of electoral illegalities (above) is adopted, some system of special elections, at least in such cases, is essential.

The Vice Presidency

The office of the Vice President and its responsibilities have been questioned from the very beginning in 1787, and some of the questions have not changed very much since that time. Today they include:

- the need for the office at all;
- the responsibilities (or absence thereof) of Vice Presidents, other than potential succession to the Presidency;
- the selection of Vice Presidential nominees almost exclusively by the Presidential nominees on the basis of balancing the ticket and/or vote-getting ability, with little regard for competence to serve as President; and
- the absence of any sufficient prior check of the individual's qualifications or disqualifications.

The consensus of the Panel is that, in order to preserve stability and continuity in the office of the Presidency, the office of Vice President should be retained.

On the question of Vice Presidential responsibilities, the Panel suggests that the Senate Select Committee consider possible alternatives, such as: giving the Vice President a voice and a vote on the floor of the Senate; giving him specific administrative responsibilities, such as making him Chairman of the Domestic Council; or continuing the situation as it is today, leaving determination of his responsibilities up to the President. Although there is no consensus on these options, the Panel's majority favors the third alternative.

On the question of the mode of selection of the Vice President, the Panel considered two alternatives: (1) providing a longer period at the nominating conventions between the nomination of the President and the beginning of consideration of the Vice President; and (2) removing Vice Presidential nomination from the convention altogether and handling the matter substantially as the Democrats did in 1972, following the withdrawal of Senator Eagleton. The Panel reached no common position on either of these questions, but suggests that the Select Committee give both consideration. (Both possibilities can of course be considered individually by the political parties themselves.)

On the question of qualifications of Vice Presidential nominees (and possibly also Presidential nominees), the Panel reached no consensus, but a majority considered that both parties should institute systematic checks of possible candidates prior to the nominating conventions.

* * *

This chapter has been fundamentally supportive of the American system of government as it was founded and as it has developed. The Panel is opposed to any change in the direction of a parliamentary system. The workings of our system obviously may be improved, and suggestions in that direction are the purpose of the chapters which follow. Such improvements should help bring about a renewal of the full respect, credibility, and confidence of the citizenry which are so essential in a democratic polity.

The primary problem discussed above is the difficulty of removing a President. The Panel feels that the difficulty arises from a fundamental misunderstanding of the sense of some old English words, including "impeachment" and "high crimes and misdemeanors." Some semantic misunderstanding may be corrected through education; on the other hand, it may require clarification or substitution of modern vocabulary, more understandable in the twentieth century, by law or Constitutional change.

The Panel reaffirms its belief in the wisdom and essentiality of a strong institutional Presidency. The intertwining of foreign and domestic policy concerns and the interaction of the whole spectrum of governmental responsibilities require the unity and leadership of a single chief executive with a perspective embracing the total public interest. Nothing in the congeries of events called Watergate negates or diminishes this requirement. Indeed, there is some danger that, in overreacting to those events, the strength and viability of Presidential leadership, both in the nation and abroad, may be damaged.

CHAPTER 2

The President, the White House, and the Executive Office

Everybody believes in democracy until he gets to the White House and then you begin to believe in dictatorship, because it's so hard to get things done. Everytime you turn around, people just resist you, and even resist their own job.[1]

AMONG the institutions of government on which Watergate has focused attention, none stands out so much as the Presidency and those staff and staff organizations surrounding the Presidency. Perhaps most controversial has been the revelation, disclosed by Presidential assistants as the rationale for their actions, that the structure which was designed to provide the President with staff assistance and advice has been gradually fashioned into an instrument of centralized control. Much of this was done in the open, building vigorously upon trends in organization of the Presidency first noted over a decade ago, and pursued in the name of efficient, effective, and responsive govern-

[1] An unidentified Kennedy aide quoted by Thomas R. Cronin " 'Everybody Believes in Democracy until He Gets to the White House': An Examination of White House–Departmental Relations," in *Papers on the Institutionalized Presidency,* a special issue of *Law and Contemporary Problems* 35, no. 3 (Summer 1970): 574.

ment. Increasingly, access to the President was restricted. The principal officers of the executive departments and agencies and the leadership (both majority and minority) of the Congress encountered more and more difficulty in seeing him. Equally important, the free flow and competition of ideas and interests were cut off.

What emerges is a picture of the centralization of power in the White House and the concomitant confusion of roles and responsibilities by placing operating authority in the hands of personal and advisory staff who make the key decisions but are shielded from public view and public access. The President needs adequate staff assistance and sufficient flexibility to serve his personal style in meeting his constitutional responsibilities; yet the full bloom of a monocratic Presidency cannot fulfill the best interests of the Republic.

Centralization of Power in the White House

Testimony and evidence presented to the Select Committee on Presidential Campaign Activities raise fundamental and disturbing questions stemming from the centralization of power in the White House: the fractionalization of Presidential power among the assistants to the President, and the division of responsibilities between the White House and the statutory agencies within the Executive Office of the President, and between the Executive Office and the executive departments and independent agencies.

According to Assistant to the President Bryce N. Harlow, "Richard Nixon is *running* the whole government from the White House." [2] The principal assistants and counselors have been converted from intimate personal advisers to the President to the equivalent of assistant presidents managing the executive establishment out of the White House. Transformation of assistants to the President into assistant presidents fundamentally alters

[2] Emmet J. Hughes, *The Living Presidency* (New York: Coward-McCann-Geoghegan, 1973), p. 344.

the role of the White House staff and has far-reaching implications for both the Executive Branch and the Congress.

The President's Committee on Administrative Management, which proposed the creation of an Executive Office of the President more than 35 years ago, emphasized that assistants to the President "would not be assistant presidents in any sense" and "would remain in the background, issue no orders, make no decisions, emit no public statements." [3] President Franklin D. Roosevelt reaffirmed this concept when he stated that his administrative assistants were "not to become in any sense Assistant Presidents, nor are they to have any authority over anybody in any department or agency." [4] Executive Order No. 8248, September 8, 1939, which is still in effect, establishing the divisions of the Executive Office of the President, directs: "In no event shall the administrative assistants be interposed between the President and the head of any department or agency, or between the President and any one of the divisions in the Executive Office of the President." [5]

The contrast between the White House Office prescribed by Executive Order No. 8248 and that described by witnesses such as H. R. Haldeman, John Ehrlichman, John Dean, L. Patrick Gray, and General Vernon Walters could not be more dramatic. The professional staff of the White House Office has grown enormously since 1939. Thomas E. Cronin, among others, has noted that:

The presidency has become a large, complex bureaucracy itself, rapidly acquiring many dubious characteristics of large bureaucracies in the process: layering, overspecialization, communication gaps, interoffice rivalries, inadequate coordination, and an impulse to become consumed with short-term operational concerns at the expense of thinking sys-

[3] U.S. President's Committee on Administrative Management, *Report of the Committee with Studies of Administrative Management in the Federal Government,* Washington, D.C., 1937, p. 5.
[4] U.S. President, "Message Transmitting First Plan on Government Reorganization," *Congressional Record* 84 (April 25, 1939): 4709.
[5] U.S. President, Executive Order, "Establishing the Divisions of the Executive Office of the President and Defining Their Functions and Duties," *Code of Federal Regulations* 3, 1938–1943, comp., p. 577.

tematically about the consequences of varying sets of policies and priorities and about important long-range problems.[6]

From analysis of the testimony and evidence, it would appear that the organization and functions of the White House office in the 1970s have been based on the following underlying premises: [7]

- The President's constitutional powers, including his inherent powers, may be delegated and may be legitimately exercised by his principal assistants acting in his name.
- The President must operate on the basis that staff come to him only when called.
- As surrogates of the President, the principal assistants must be self-starters in policy making because, according to John Ehrlichman, "In the Nixon White House there is no one else who is going to have the time to supervise, make assignments, decide what should be looked into. . . . It would be impossible for the President, or any one person in his behalf, to keep informed of everything being done by the staff, even in areas of major current interest or concern."
- Department and agency heads must obey orders from White House staff even in those areas where statutory powers are vested in them, and they are legally accountable for the actions taken.
- Agency heads should understand that when a request comes from the White House, they must accomplish it without questioning the merits. Even suggestions from the President's principal assistants are to be construed as orders coming directly from the Oval Office.
- The bureaucracy is engaged in a kind of guerrilla warfare against the President. The lack of key Republican bureaucrats or "RN supporters" at high levels precludes the initiation of policies which would be proper and politically advantageous.
- The President is both the nation's Chief Executive and the leader of his political party. Members of the White House staff are properly assigned "political duties" and are expected to supply the President with the information he needs and wants about issues, supporters, opponents, and every other political subject known to man.

In the light of events, it seems evident that recognition of the meaning, impact, and results of these premises came only after their application clearly led to trouble. In some ways the reorganization announced by President Nixon on January 5, 1973,

[6] Thomas E. Cronin, "The Swelling of the Presidency," *Saturday Review of the Society* 1, no. 1 (February 1973): 34.

[7] From testimony and exhibits presented or furnished to the Senate Select Committee on Presidential Campaign Activities by John W. Dean III (June 27, 1973), John Ehrlichman (July 24, 1973), and L. Patrick Gray (August 3, 1973).

"to revitalize and streamline the federal government in preparation for America's third century," represents an attempt to formalize the transfer of control of Executive Branch policies and programs from responsible agency heads to the White House.[8] The reorganization established a type of structure with a rigid hierarchy in which:

· Access to the President was limited to five assistants to the President (a more accurate description would be assistant presidents).
· Four of the five assistants to the President would act as Presidential surrogates with responsibility to integrate and unify policies and operations in the following areas: domestic affairs, foreign affairs, executive management, and economic affairs.
· Access to the assistant to the President (Domestic Affairs) would be limited, with some exceptions, to three counsellors (to be housed in the Executive Office Building) for Human Resources (Secretary of HEW), Natural Resources (Secretary of Agriculture) and Community Development (Secretary of HUD). The counsellors would have "coordinating responsibilities" within their assigned areas, which were interpreted to include budget review, control of key appointments, and policy direction on legislation, among other functions.

The following diagram illustrates the concept:

THE PRESIDENT

Ass't. for White House Operations (Haldeman)	Ass't. for Management (Ash)	Ass't. for Domestic Affairs (Ehrlichman)	Ass't. for Economic Affairs (Schultz)	Ass't. for National Security Affairs (Kissinger)

Presidential Counsellors for

Natural Resources (Butz)	Human Resources (Weinberger)	Community Development (Lynn)

Domestic Departments and Agencies	Defense, State, CIA

This reorganization was formally abandoned in May 1973.

[8] *Weekly Compilation of Presidential Documents* 9, no. 3 (January 8, 1973): 5–10.

Among the principal questions which need to be addressed are the following:

1. Are the President's constitutional powers divisible? When and under what circumstances may the President delegate powers to subordinates?
2. Is there a valid distinction between Presidential advisers and White House staff members acting as federal officers making decisions or issuing directions on their own authority? Should the privileges and immunities traditionally extended to Presidential advisers also apply to others employed by the White House or to those employed elsewhere who are designated as White House counsellors?
3. Are the roles of the President as chief executive, party leader, and political candidate so intertwined that they cannot be clearly demarked? What are the limits, if any, which should be applied to White House staff activities which have as their prime objective promotion of the President's political interests?
4. Should the President have unrestricted authority to assign to department and agency heads a dual responsibility as Presidential assistant and deprive the agencies of effective full-time leadership?

Delegation of Presidential Powers

Alexander Hamilton wrote in *The Federalist:*

. . . the plurality of the executive tends to deprive the people of the two greatest securities they can have for the faithful exercise of any delegated power; first, the restraints of public opinion, which lose their efficacy as well on account of the division of the censure attendant on bad measures among a number, as on account of the uncertainty on whom it ought to fall; and, secondly, the opportunity of discovering with facility and clearness the misconduct of the persons they trust, in order either to their removal from office, or to their actual punishment in cases which admit of it.[9]

Our Constitutional system is based on the principle of the unity of executive power and the premise that "the executive power is more easily confined when it is one." [10] The concept of assistant presidents sharing by delegation or otherwise in the powers ex-

[9] Alexander Hamilton, *The Federalist,* ed. Jacob E. Cooke (Middletown, Conn.: Wesleyan University Press, 1961), no. 70, pp. 477–478.
[10] Quoted by Hamilton from Jean Louis Delolme, ibid., p. 479.

clusively vested in the President by the Constitution cannot be reconciled with this principle.

Under the provisions of the McCormack Act of 1951, the President is implicitly debarred from making formal delegations to members of the White House staff.[11] The McCormack Act authorizes the President to delegate statutory functions, without relieving himself of responsibility for their proper performance, to the head of a department or agency, or any other official of the Executive Branch whose appointment is confirmed by the Senate. Such delegations must be in writing and published in the *Federal Register*. Members of the White House staff are not heads of departments or agencies, and they are not appointed with Senate confirmation.

The authority conferred by the McCormack Act was intended to apply to "routine" functions and provides no authority to delegate Constitutional functions. The legislative history indicates that the bill was designed to relieve the President from performing functions which "do not have any reasonable claim upon his time or attention." [12]

White House staff have no legitimate power of their own, and whenever they exercise power they are doing so in the name of the President. The Watergate record graphically demonstrates the disastrous consequences of allowing the Presidency to speak with many voices. On the other hand, the President should be relieved of ministerial, routine functions by delegation to his assistants.

If, as claimed by Ehrlichman and Haldeman, the principal assistants to the President act under delegated Presidential authority, then it is essential that the limits of their authority and responsibility be expressly defined by executive order.

The McCormack Act should be modified to permit delegations of routine functions to White House staff, and the Act should be supplemented by the enactment of legislation requiring the President to state and currently publish in the Federal Register: *(1) a description of the organization of the White House office and*

11 65 stat., 712; 3 U.S.C. (1970), pp. 301–303.
12 1951 *U.S. Code Congressional Service*, pp. 2931–2937.

the functions assigned to the various units within the office; (2) the names and duties of any personnel detailed to the White House office with or without reimbursement; (3) delegations of authority within the office; and (4) the titles of officials to whom delegations have been made.

Distinction between Advisers and Officers

The distinction between intimate personal advisers to the President and those exercising executive power on behalf of the President is crucial. Difficulties occur when the privileges and immunities traditionally extended to advisers are assumed to apply to everyone in the White House, regardless of position.

The petitioner's brief in the case of *Nixon* v. *Sirica and Cox* stresses the unique attributes of the President which distinguish him from other officers of the Executive Branch. The argument for executive privilege is based on the need to preserve the confidentiality of communications between the President and his "intimate advisers." Nowhere is the argument made that the unique attributes of the President are shared by White House staff acting as Presidential surrogates.

If members of the White House staff are to continue to act in a capacity other than that of personal advisers and staff assistants to the President, and are authorized to issue "orders" on the President's behalf or on their own authority, then modifications are called for in present interpretations of executive privilege. Indeed, the question may well be raised as to whether assistants to the President who perform executive functions occupy "offices" subject to Article II, Section 2, of the Constitution, which provides that the President "shall nominate, and by and with the advice and consent of the Senate, shall appoint . . . all other officers of the United States, whose appointments are not herein otherwise provided for, and which shall be established by law. But the Congress may by law vest the appointment of such inferior officers as they think proper in the President alone, in the courts of law, or in the heads of departments." Principal White House assistants with the equivalent of cabinet rank and acting as supercabinet officers would not appear to belong in the category of inferior officers.

Maintenance of a distinction between advisers and staff assistants and administrators is also in the President's own interest. A President does not protect himself against the special pleaders in the Executive Branch by transferring them to his own household. Staff with specific operating or investigatory responsibilities and intent on covering up their mistakes are not likely to provide the confidential, independent, objective advice and assistance which any President requires. Furthermore, the President cannot disavow acts by White House aides even when acting on their own. All of the mistakes become the President's mistakes.

Legislation should be enacted along the lines of Executive Order No. 8248 prohibiting assistants to the President from issuing orders and interposing themselves between the President and the head of any department or agency or any one of the divisions in the Executive Office of the President. The legislation should include appropriate sanctions to be applied to both the initiator of such illegal orders and those officials who accept and carry them out. This would not preclude liaison activities, collection of information on the President's behalf, or exercise of routine delegated functions.

The tendency has been for Presidents to assign to White House assistants or to Executive Office agencies certain types of functions which do not fit neatly into the existing Executive Branch structure. This problem can to some degree be ameliorated by Executive Branch reorganization, but there will always be significant activities which cut across established jurisdictional lines and require a high degree of interagency collaboration. Institutional arrangements should be devised which are specifically adapted to the performance of such temporary functions as the development and presentation to the Congress of major legislative proposals of a multiagency character (e.g., energy, poverty, trade), organization of emergency programs, and short-term coordinating and expediting assignments. At present the President is often confronted with a choice among a number of equally objectionable organizational options.

To increase the options now available to the President, he should be authorized by law to create a limited number of temporary offices of secretarial rank outside the Executive Office of the President to which he could delegate ad hoc assignments, including coordination of programs involving a high degree of interdepartmental collaboration. These would in no sense be supersecretaries, and assignments generally would be of short duration. Such "secretaries without portfolio" would be appointed by the President, by and with the advice and consent of the Senate, and be paid salaries at the level of the heads of executive departments. Their duties and authorities would be set forth in executive orders. They would be expected to testify before the Congress and would be accountable to the Congress and the President within the areas of their assigned responsibilities.[13]

Political Activities

A candidate does not cease to be a politician upon election to the Presidency. Indeed, a first-term President remains a potential political candidate. Politics has a legitimate role in the democratic process, and few Presidential actions do not have political implications. Indeed, the political fortunes of the President and his political party are directly influenced by the success or failure of his domestic and foreign policies and programs.

While the line between political and nonpolitical may at times be shadowy, there are and must be limits on political partisanship. As with other public employees, the first duty of White House employees is to the nation, not to the incumbent President or to a political party. Public office, public authorities, and public funds are not to be exploited to damage or embarrass one's political opponents. Such activities by the White House staff as

[13] Benjamin V. Cohen, former assistant to President Franklin D. Roosevelt, supports and discusses this concept in "The Presidency as I Have Seen It," Hughes, *The Living Presidency*, p. 323.

solicitation and custody of campaign contributions, political intelligence and espionage, attempts to use authorities of agencies for purely partisan political ends, and political "dirty tricks" demonstrably do not fall within the limits of permissible political activities. For the White House to engage in such activities degrades the institution of the Presidency. As a political candidate, a President should look outside the White House and the government for legitimate campaign assistance.

Ultimately it is the President who must set the tone and example for his immediate staff. Reduction in the size of the present White House establishment would help to reduce the risk of improper activities. If staff are fully occupied in the conduct of the public business, they will not have time for extracurricular activities. The record would indicate that substantially more personnel are now employed in the White House than are necessary to support and assist the President in carrying out his *Presidential* duties. This fact was acknowledged by President Nixon in his message transmitting Reorganization Plan No. 1 of 1973, in which he stated that a "leaner and less diffuse Presidential staff structure . . . would enhance the President's ability to do his job."

The White House staff should be limited in number to perhaps 15 top personal aides to the President and 50 supporting professional employees, the latter to be subject to the political activity limitations of the Hatch Act. (This limit on number of staff would not include domestic and housekeeping staff, White House police, or clerks and secretaries.)

Dual Offices

Special problems are created by the designation of cabinet officers and the Director of the Office of Management and Budget as assistants and counsellors to the President. Such "two-hatted" arrangements are by no means novel, but the extent and manner of their use by President Nixon are unprecedented.

Two-hatted arrangements are disadvantageous for the President, the Congress, and the individual officers and the agencies for which they are responsible. The two jobs are not compatible. A cabinet member cannot at one and the same time be an effective advocate of departmental positions and a neutral agent of the President. The dual roles have built-in conflicts of interest. Both jobs require full-time attention if they are to be performed well. When the assistants to the President and counsellors occupy offices in the White House or the Executive Office Building, they inevitably become isolated from their agencies, which are cast adrift without a full-time head.

The individual officer will find it difficult to draw a hard and fast line between what he does in his White House capacity and his other duties. This can prove a source of embarrassment. He may be faced with the dilemma of either denying to the Congress information to which it is rightly entitled, or breaching privileged communications with the President. The blurring of responsibilities provokes challenges to the immunities traditionally accorded Presidential aides.

The Congress is faced with a form of Executive Branch shell game. It is almost impossible to ascertain who is responsible for a given decision—the President himself, an assistant to the President, a counsellor, or the agency head who is legally responsible and accountable both to the President and the Congress.

Legislation should be enacted prohibiting the heads of executive departments and agencies from serving as assistants or counsellors to the President.

Obviously this legislation could in no way impair the President's Constitutional right to seek the opinions and advice of Executive Branch officials.

Restoring the Staff Role to the Executive Office

The President's Committee on Administrative Management in 1937 conceived of an Executive Office of the President as consisting of several permanent agencies staffed mainly by career employees, which would serve the interests of the Presidency as an institution, and a handful of White House assistants, who would serve the personal, political interests of the President. Given the changes of 37 years and the revelations of Watergate, is it possible to maintain a distinction between those agencies in the Executive Office of the President serving the interests of the Presidency as an institution and staff serving the personal, political interests of the President? Has the ability of the Executive Office of the President to tend to the President's business been impaired by the inclusion of special staff and operating agencies?

The rapid expansion of the White House office has been accompanied not only by a decline in the influence of Cabinet members, but also by a confusion in the roles and responsibilities among the staff in the Executive Office of the President serving the institutional responsibilities of the President and those serving him in a personal, political capacity. In his message transmitting Reorganization Plan No. 2 of 1970, President Nixon noted "a tendency to enlarge the immediate White House staff —that is, the President's personal staff, as distinct from the institutional structure. . . . This has blurred the distinction between personal staff and management institutions." [14]

Steps should be taken to reemphasize the essential distinction between staff serving the President and staff serving the Presidency by discontinuing the designation of the Director of the Office of Management and Budget as assistant to the President, and amending Reorganization Plan No. 2 of 1970 to provide a full-time Director of the Domestic Council in lieu of an assistant to the President designated to perform that function.

The decline of the institutional agencies in the Executive Office of the President results at least in part from the proliferation of agencies within the Executive Office. The Congress must assume much of the responsibility for those developments. Congressional pressure to provide visibility and enhanced status for favored programs has diverted the Executive Office of the President from its exclusive concern with *Presidential* business by establishing

[14] Message from the President of the United States transmitting Reorganization Plan No. 2 of 1970, *Hearing Before the Subcommittee on Executive Reorganization and Government Research,* Committee on Government Operations, United States Senate 91/2, May 8, 1970, p. 4.

within it such agencies as the Office of Economic Opportunity, National Council on Marine Resources and Engineering Development, and Council on Environmental Quality, and by providing that certain appropriations are made to the President—a technical absurdity. The number of employees in the Executive Office of the President grew by nearly 60 percent from 1,403 in 1955 to 2,236 in 1972, creating unnecessarily difficult coordination and management problems within the President's own institutional machinery.[15]

The Executive Office of the President should be streamlined by removing from it all agencies with special purposes which are not primarily concerned with providing staff assistance to the President.

The present structure of the Executive Office of the President does not reflect the intention of either the Committee on Administrative Management or the first Hoover Commission that no institutional resources be provided within the Office other than those which the President found essential to advise and assist him in carrying out functions which he could not delegate.

The purpose of the staff service in the President's office is not to assume operating functions or to duplicate responsibilities of the operating departments. These services should exist only to give the President the greatest possible information on the activities of the Government as a whole, and to enable him to direct the policies of the departments and agencies.[16]

The principle remains valid today. President Nixon, in his message transmitting Reorganization Plan No. 1 of 1973 to the Congress, called for a

sharp reduction in the overall size of the Executive Office of the President and a reorientation of that office back to its original mission as a staff for top-level policy formation and monitoring of policy execution in broad functional areas. The Executive Office of the President

[15] *A Report on the Growth of the Executive Office of the President, 1953–1973,* Committee on Post Office and Civil Service, U.S. House of Representatives, 92/2, April 24, 1972, Table IV, p. 5.

[16] The U.S. Commission on Organization of the Executive Branch of Government, *General Management of the Executive Branch, A Report to the Congress* (Washington, D.C.: U.S. Government Printing Office, 1949), pp. 14–15.

should no longer be encumbered with the task of managing or administering programs which can be run more effectively by the departments and agencies.

The functions of the Executive Office should be to assist the President, not to be the general manager of the Executive Branch.

The Office of Management and Budget

In any effort to restore the staff role to the Executive Office of the President, the Office of Management and Budget (OMB) must receive particular attention. To comprehend this imperative, it is necessary to compare the philosophy which guided the Bureau of Budget (BOB) with that of the current philosophy of the OMB.

Historically, the BOB was the principal staff arm of the President in: (1) formulating and administering national budgets and advising on fiscal policy; (2) coordinating and reconciling legislative programs and proposals; (3) assisting in resolution of interagency policy and administrative disputes; (4) designing management systems and reorganization proposals; and (5) providing an institutional memory with regard to federal policies and programs. These central functions traditionally were performed by the institutional career staff of the BOB under the direction of the Director and Deputy Director as Presidential appointees. Budget Directors did not seek or assume, or attempt to have delegated to them, decision-making powers which were uniquely the province of the President or agency heads. The Bureau functioned as a catalyst and coordinator. It was perceived generally as a center of expertise and a source of trusted advice. Its staff was regarded as an expert, nonpartisan, professional corps of dedicated public servants who were in close communication with the President through the Budget Director and designated White House staff. They were active participants in the policy formation process, but they served in a counseling and advisory role. Decisions in all except routine matters were made by the President and sometimes announced in his name by the Director.

In 1970 the President transmitted Reorganization Plan No. 2, which established the Domestic Council and redesignated the Bureau of the Budget as the Office of Management and Budget. The latter change was designed primarily to emphasize the management functions of the office. For several years prior to 1970 a significant growth in the White House staff, formalized by the establishment of the Domestic Council, had begun to affect the relationship between the White House and BOB/OMB. This new relationship had the effect of transferring to the White House staff the primary program advisory role previously assigned to BOB, and much of the hitherto assigned responsibility for deciding the basic thrust of program coordination.

Further profound changes were made in the role and functioning of OMB subsequent to the 1972 election: (1) assignment of political officers as the primary senior level of the OMB staff; and (2) an additional internal reorganization designed to increase the influence of noninstitutional staff through the appointment of management associates, appointed on a noncareer basis and outside the career tradition. There was a clear intent, through these changes, to assure complete responsiveness by OMB staff to White House guidance.

The emergence of a powerful White House staff which has progressively assumed the role of speaking for the President has seriously diminished the responsibilities of the career, professional staff of OMB and its capacity to provide the kind of objective and expert counsel to the President which characterized earlier operations, when career division chiefs had responsibility to be the eyes and ears on agency programs and policies. Increasingly, the White House staff has come to parallel and, to some extent, duplicate and supersede the OMB staff. This situation is further aggravated by the interposition of a layer of political executives between the career staff of OMB and the Director and the President. As a consequence, the OMB institutional staff is largely isolated from the decision-making process in policy and program formulation. The removal of the Director's office to the White House and his service as counsellor to the President has limited the access of the staff both to the Director of OMB and to the President. Insulated from the White House policy center and subordinated to a group of basically business (rather than government-trained) executives who fill the top posts in OMB, the career staff now serves primarily a ministerial role.

The guiding philosophy of OMB should be that it is a staff arm to the President, providing him with advice and counsel, but with basic decision-making responsibility reserved for and exercised by the President.

The style of management possibly well-suited to the needs of some private corporations, now currently in vogue in OMB, is not automatically or necessarily well-adapted to the public service, and should be thoroughly reexamined and reconsidered.

Open channels of communications between the career, professional staff, and the top political staff of OMB should be reestablished and maintained; senior institutional staff of OMB should play a strong participative and contributory role in the decision processes of the Executive Office of the President.

Policy and final decision making in the Executive Branch should be the product of interaction between the President and agency heads.

Program management should be an agency responsibility.

* * *

Few would dispute the view expressed by the House Committee on Government Operations that "as a practical proposition, the President cannot be compelled to utilize a policy making and advisory apparatus against his own preferences. Furthermore, the President should have considerable latitude in determining the composition of the Executive Office." [17] The President ought to have the capability to adapt the Executive Office machinery to his perceived needs, but he cannot be permitted in the process to ignore the general welfare of the nation, or the needs of future Presidents and the Congress. If any lesson has been learned from Watergate, it should be the significant implications of White House and Executive Office organization and staffing for the effective functioning of our Constitutional system.

[17] U.S. Congress, House Committee on Government Operations, Reorganization Plan No. 1 of 1973. House Report 93–106, p. 18.

CHAPTER 3

The Chief Executive and the Executive Branch

There shouldn't be a lot of leeway in following the President's policies. It should be like a corporation, where the executive vice presidents (the Cabinet officers) are tied closely to the chief executive, or to put it in extreme terms, when he says jump, they only ask how high. —John Ehrlichman [1]

THE PROPER relationship between the President and his office and the departments and agencies which comprise the Executive Branch has been a source of debate as old as the Republic itself. George Washington certainly supposed, and so did the officials of his Administration, that he could summon any of them for a review of any pending matter and instruct them as to its disposition. This derived from, among other things, his power to "require the opinion, in writing, of the principal officer in each of the executive departments upon any subject relating to the duties of their respective offices. . . ." [2] But in a contrary direction, Congress, in the Treasury Department Act of 1789, established five distinct offices, each with distinct responsibilities. Subsequent practice, except for emergency legislation, has

[1] In an interview published in *The Washington Post*, August 24, 1972.
[2] The Constitution, Art. II, Sect. 2.

44

been to vest statutory powers in bureaus or agencies, or their heads, rather than in the President.

Andrew Jackson, bucking this trend, dismissed two Treasury Secretaries in order to get a third who would withdraw federal deposits from the Bank of the United States, and so set a precedent for President Nixon's recent firings in the Justice Department. But two Supreme Court decisions a century apart (*Kendall v. U.S.* in 1838 and *Humphrey's Executor v. U.S.* in 1935) have given emphatic approval to the doctrine that Congress can vest powers in agencies or officials, to be exercised on their own responsibility, except for some military and foreign affairs and some domestic matters of high policy.

Accustomed patterns of relationships between the President and different agencies have developed over the years in practice as well as law. First are the 11 departments which constitute the Cabinet agencies and which are generally regarded as more prestigious than the others. As one perceptive scholar noted a few years ago, the Cabinet is itself divided into two groups: an Inner Cabinet, consisting of the Departments of State, Defense, Treasury, and Justice, and an Outer Cabinet which includes all the others.[3] The Secretaries of the Inner Cabinet have customarily been men for whom the President has high personal respect and confidence. He calls upon them frequently for counsel, sometimes in fields outside their departmental responsibility. He relies heavily upon them in both the development and execution of policy. It is probably not an accident that the Secretaries of State, War (Defense), and Treasury, and the Attorney General, were the original Cabinet, established by the First Congress in its first year.

Relationships with the Outer Cabinet are quite different. All operate primarily in domestic fields, and their Secretaries are usually viewed, and view themselves, as representatives of their Departments and the clienteles and interests which they serve. They are called upon by the President much less frequently and sometimes find access to him nearly impossible. Their relation-

[3] Thomas R. Cronin, " 'Everybody Believes in Democracy until He Gets to the White House': An Examination of White House–Departmental Relations," in *Law and Contemporary Problems* 35, no. 3 (Summer 1970): 610.

ship with the President—and his staff—is frequently more nearly that of adversary than confidant.

Beyond the Cabinet departments are a large number of executive agencies: the Veterans Administration, the General Services Administration, the TVA, the Atomic Energy Commission, NASA, and many others. They have widely varying degrees of autonomy and independence, legal and operational, but on the average it is probably true that they have still fewer contacts with the White House, unless their field of activity suddenly becomes controversial. And beyond these are the independent regulatory commissions, the substantive activities of which are intended by law to be insulated from White House or other outside penetration.

All the Presidents of the last 40 years, and many of their staffs, have at one time or another expressed exasperation at their inability to "get things done" or to get effective response to their policy preferences. In the case of Inner Cabinet departments, the criticism is usually directed to elements of the department below the Secretary. And the feeling has been returned by agency heads and their subordinates, protesting lack of access, directives on matters on which they were not consulted, and "meddling" by White House staff. Here, as in many other features of American government, are built-in tensions which are probably unavoidable in the system.

The Corporate State

It now appears that recent administrations have been moving toward fundamental changes in that system which, if fully achieved, would have substituted a governmental philosophy foreign to the concepts of the framers of the Constitution and the concepts of most Americans since then. The U.S. Government would be run like a corporation—or at least a popular view of the corporate model—with all powers concentrated at the top and exercised through appointees in the President's office and loyal followers placed in crucial positions in the various agencies of the

Executive Branch. This would have effectually destroyed public accountability except for the President himself.

No one can guess how close the American government would be to this closed hierarchal model had not the Watergate exposures halted the advance toward it—at least temporarily.

The Panel believes that the Select Committee should emphatically reaffirm that:

- *The President is recognized to have general authority and responsibility over the agencies in the Executive Branch, subject to restrictions on such authority in law, and over the nature and direction of public policy within the framework of law;*
- *Virtually all the executive agencies were established by law, duly passed by the Congress subject to Presidential veto, or by Reorganization Plan, initiated by the President, subject to Congressional veto; with a few exceptions, all received their powers and responsibilities in the same ways, not by delegation; and, for the most part, their ability to operate is annually renewed through appropriations, passed by Congress, subject to Presidential veto;*
- *The top leaders in almost all the agencies are appointed by the President, subject to confirmation by the Senate;*
- *The heads of the agencies are therefore responsible, in different ways, to both the President and the Congress.*

More specific means of striking a reasonable balance between the Presidency and the agencies are suggested in other chapters, specifically those which deal with the Executive Office, Congressional oversight, and the public service.

One of the justifications for White House and Executive Office intervention in both operational and policy matters is that so many of the social problems faced by the government cross agency lines and cannot be contained within the confines of any single agency. To alleviate this problem on domestic matters, a number of reorganization proposals have been made over the years which would seek to provide mechanisms whereby these interdepartmental matters could be adequately handled below the level of the President.

The Panel endorses the aims of departmental reorganization to bring together, under single secretaries, activities of bureaus and services which are closely related and interdependent, with the understanding that the responsibilities for developing and executing policies in their assigned fields, subject to Presidential and Congressional review and approval, be vested in the appropriate secretaries.

The Use of Agency Powers for Political Purposes

One of the abiding principles of democratic government is that of *equal treatment under the laws.* Ideally, individuals, corporations, foundations, universities, or communities eligible for consideration for either reward or penalty should be judged according to common and relevant criteria without regard to race, sex, religion, or *political beliefs and activities.* In those areas where discretion to reward or punish has been delegated to executive agencies, there is a long and deplorable history of violation of this principle. Recently, much of this activity was apparently stimulated and directed from the White House or its adjunct, the Committee to Reelect the President. Most of it was to have been carried out through the powers of the operating agencies of the government, although agency resistance provoked the White House and the Committee to act on their own on some occasions.

The negative aspect of preferential treatment for partisan purposes was illustrated by the "enemy list," the testimony of Patrick Buchanan, and the efforts—largely unsuccessful, it appears—to punish opponents through the Internal Revenue Service. It was epitomized in John Dean's memorandum to John Ehrlichman of August 16, 1971, which reads:

This memorandum addresses the matter of how we can maximize the fact of our incumbency in dealing with persons known to be active in their opposition to our Administration. Stated a bit more bluntly—how we can use the available federal machinery to screw our political enemies. . . .

In brief, the system would work as follows:

· Key members of the staff (e.g., Colson, Dent, Flanigan [sic] Buchanan) should be requested to inform us as to who [sic] they feel we should be giving a hard time.

- The project coordinator should then determine what sorts of dealings these individuals have with the federal government and how we can best screw them (e.g., grant availability, federal contracts, litigation, prosecution, etc.).
- The project coordinator should then have access to and the full support of the top officials of the agency or department in proceeding to deal with the individual.

It is needless to reiterate here the threats and efforts so fully documented in the Select Committee's testimony.

The positive aspect of preferential treatment for partisan purposes was the granting or the implied promise of such treatment to individuals and organizations on provision that they would contribute to, or otherwise support, the incumbent Administration and its continuance; or the implied intimidation if they did not contribute and support. The extent and damage of this kind of behavior are not yet known. Certainly, the available evidence suggests that they may have been great: the ITT case, the milk cooperative case, and the Small Business Administration cases; the fund-raising activities of Mr. Kalmbach; the curious financial relations of Howard Hughes and the Nixon Administration—all of these and perhaps others not yet disclosed appear to exemplify the utilization of governmental powers for partisan purposes.

It would be impossible, through legislation alone, to guarantee against these kinds of political infractions of the principle of equal treatment under the laws. The essential ingredients lie in the integrity and the sense of public interest of the President and his appointees; these can hardly be assured through legislation. But it is possible that more effective disincentives and safeguards against the abuse of government powers could be built into the system.

The Panel recommends that generally the President, his staff, the Executive Office, and the heads of executive agencies refrain from participating in cases involving individuals or specific institutions, but rather concentrate on the policies and criteria governing such cases and rely upon the operating agencies to apply them.

In those few cases where top executive involvement is re-

quired, the Panel recommends that a written record be maintained.

And finally it urges that any infractions of such guidance be investigated and prosecuted by the Permanent Special Prosecutor, proposed in the following chapter.

Law-Enforcement and Intelligence Agencies

The most alarming of the Watergate disclosures was the attempted use and misuse of the law-enforcement and intelligence agencies by the White House against supposed "enemies" of the Administration, most of whom were American citizens. Americans may be everlastingly grateful and proud that most of these political pressures from above in the hierarchy were resisted from within. The Watergate hearings are replete with examples:

· The plan to involve the investigative agencies in domestic espionage involving surreptitious entry, electronic surveillance, opening of mail, and use of undercover agents on college campuses, aborted by the disapproval of the Director of the FBI;
· Wiretaps on a number of government officials and newsmen considered unfriendly to the Administration;
· Destruction of documents by the Acting Director of the FBI on instructions of a principal White House aide;
· The assistance, to a limited extent, of the CIA in the burglary of Ellsberg's psychiatrist's office;
· The White House effort to persuade the CIA to provide a national security cloak over the laundering of campaign contributions through Mexico;
· White House efforts to persuade the CIA to provide bail and support for the Watergate conspirators;
· The White House enemies list, provided to the Internal Revenue Service for tax audits and denial or scrutiny of tax exemptions;
· The White House compromising of the FBI investigation of the Watergate burglary by having the President's counsel sit in on all interviews with White House employees.

The examples might go on indefinitely, and maybe they will continue to do so. A fairly clear pattern so far emerging is that of efforts by the White House to use the resources of the law-enforcement and intelligence agencies for self-serving or political purposes.

The protection of the rights, liberties, and privacy of American citizens requires the continuing scrutiny of governmental operations in these fields by the Congress of the United States.

The Panel recommends that appropriate committees of Congress give special attention and oversight to the agencies like the FBI, the CIA, the IRS, and others whose activities might infringe upon the liberties and the privacy of United States citizens. Legislation should be enacted which would specifically prohibit conduct like that disclosed on the part of top officials of the FBI and the CIA. Such legislation should, additionally, limit access to the records of these agencies, plus the IRS, except for law-enforcement purposes. Further, Congress should specifically prohibit the White House from conducting intelligence activities itself.

Despite such legal strictures on the activities of these key agencies as may be enacted, the most profound influence on the conduct of organizations is the leadership model which the agency head projects. Professionalism and adherence to high legal and ethical standards will quickly disappear in an atmosphere that does not encourage maintenance of such standards. Hence, it is essential that heads of agencies such as the CIA, the FBI, and the IRS be selected with the greatest care, with overriding attention paid to the professional and ethical qualifications of the nominees. In Chapter 5 on the Public Service, the Panel recommends that political appointees to specialized positions be reviewed by a select group of experts to determine their fitness for office.

The Panel recommends that its proposal for an expert panel review of the qualifications of certain Presidential appointees be specifically applied to nominations for the heads of the CIA, the FBI, and the IRS.

* * *

This chapter has attempted to emphasize the necessarily pluralistic nature of the federal establishment. Whereas the President properly exercises overall responsibilities for the operations and policies of the Executive Branch, the individual departments and agencies must enjoy a degree of independence from the White House, consistent with their dual allegiance to the Presidency

51

and to the Congress which participated in creating them and to which they must look for both sustenance and program approval. Finally, the intelligence and investigative agencies are particularly subject to criticism resulting from excessive White House direction, and must be afforded an even greater degree of freedom from highly centralized control in the Executive Branch.

CHAPTER 4

The Attorney General
and the
Department of Justice

> *The Office I hold is not properly po-*
> *litical, but strictly legal; and it is my*
> *duty, above all other ministers of State*
> *to uphold the Law and to resist all en-*
> *croachments, from whatever quarter,*
> *of mere will and power.*
> —Attorney General Edward Bates
> (1862–1864)

THE OFFICE of Attorney General, together with those of United States Attorneys in all of the judicial districts, was created by the First Congress in 1789. The Attorney General was to represent the nation in cases before the Supreme Court and was further called upon to provide legal advice to the President and to the heads of the departments. The U.S. Attorneys were made responsible for representing the interests of the United States in all criminal and civil proceedings in their districts. Problems arising in enforcement of the Reconstruction Acts led to the establishment, in 1870, of the Department of Justice under the Attorney General. Since that time the Department has been given an increasing and increasingly variegated range of functions. Today, with its nearly $2 billion budget and 48,000 employees, it combines the world's largest law office with a number of other activities directly or indirectly related to the criminal justice system. The scope of the Department is summarized below.

1. The Attorney General, a cabinet member, is the lawyer for the President and is charged with the duty of rendering advice and opinions, upon request, to the President and heads of other executive departments. While each of the other departments has its own counsel, the Attorney General is the chief law officer of the United States. Should there arise differences in view on a subject among the myriad lawyers within the Executive Branch, it is quite clear that the Attorney General has the power to resolve such disputes. One has always supposed that the legal advice rendered by the Attorney General to the President and the Executive Branch heads related to the performance of their official duties and not to their personal or political affairs.

2. The Attorney General administers the Department of Justice and represents the United States of America in all court proceedings, from those in the Supreme Court of the United States to those pending in the lowliest state courts where an agency of the federal government may have some interest or where a federal question may arise.

3. The Department is charged with the prosecution of violations of federal laws as part of the Executive Branch's duty to enforce federal statutes. This function is carried out largely through 94 U.S. Attorneys who, with their staffs, are located in each of the country's federal judicial districts. Although the activities of the U.S. Attorneys are centrally directed and managed within the Department of Justice, the U.S. Attorneys (Presidential appointees) and their staffs (frequently political appointees) and the U.S. Marshals reflect varying degrees of professional competence and allegiance to the Department. As a consequence, these "local" representatives of the Department of Justice are not always responsive to departmental policies. The prosecutorial responsibility to investigate Watergate was located, at the outset, in the U.S. Attorney's Office for the District of Columbia. Apparently the specter of the Executive Branch investigating itself was considered, because the White House made a point of publicizing that the investigation was headed up by a Democrat. Supervision of the investigation remained with the Criminal Division of the Justice Department, whose head kept the White House

advised. The Select Committee testimony of Dean, Gray, and others highlighted the enormity of the conflict of interest that existed when the Department attempted to conduct an inquiry into Watergate.

4. It is traditional for the Attorney General, through his deputy, to screen and recommend to the President all appointments to the federal judiciary—new district court judges, judges on the various Courts of Appeal, and justices on the Supreme Court of the United States.

5. The Department administers several grant and operating programs, such as the Law Enforcement Assistance Administration, the federal prison system, and the federal parole boards, and functions as the recommending agency for Presidential pardons.

6. In theory, at least, the Attorney General supervises and controls the Federal Bureau of Investigation as an aid in the discharge of his statutory responsibilities. Historically, however, there has been little supervision and control.

7. The Department comments on legal issues in proposed legislation being considered by Congress.

The Department now consists of: the Attorney General, his Deputy, the Solicitor General, and a number of Assistant Attorneys General in charge of divisions in different aspects of the law (such as criminal, civil, and antitrust); the central staffs of the divisions, most of which are staffed with well-qualified lawyers in a de facto career system (though not under civil service); the U.S. Attorneys and their staffs in the field; and a number of operating bureaus and services which are in varying degrees supervised by or autonomous from the central Department.

Many of these elements of the Department were (and in some cases still are) involved in Watergate in various ways and at different times. Many of the difficulties highlighted in the Watergate investigations stem from two major sources, both of which have a long history: (1) the politicization of certain parts and processes of the Department; and (2) the accretion of a large variety of responsibilities that are not consistent with the conduct of a government law office.

Politics and Justice

Since World War II, it seems to have become nearly standard practice for the President to appoint as Attorney General one of the principal leaders of the political campaign election in which he was elected. Every President since then, except President Johnson, has done so. President Truman named McGrath, President Eisenhower named Brownell, President Kennedy named his brother Robert, and President Nixon named Mitchell. Several of these Attorneys General continued, after appointment, to serve as principal political advisers to the President, as well as providing him with his legal counsel, and John Mitchell resigned the post to head the Nixon reelection campaign.

By historical tradition, and with certain notable exceptions, the top Presidential appointees in the headquarters of the Department have been highly respected representatives of the legal profession. Recent trends to appoint unsuccessful candidates for political office to a number of the Assistant Attorney General posts have undermined the tradition. The evidence of Watergate suggests that politicization in the Justice Department has not served the national interest.

While Presidential appointees to other Departments have not come from monasteries, the partisan orientation in recent years of those responsible for supervising the nation's legal affairs seems extreme. Its 94 principal field representatives, the U.S. Attorneys, are appointed by the President with senatorial consent for four-year terms, and the custom of senatorial courtesy inevitably involves such appointees in partisan politics. The impact of politics is intensified by the fact that some 1,200 Assistant U.S. Attorneys are appointed by the U.S. Attorneys, frequently on the basis of party patronage. The U.S. Marshals are likewise typically appointed on a patronage basis. Finally, the suggestions of federal justices for appointments and promotions from district courts to the Supreme Court have devolved upon the Department, and thus both the Department and the judiciary are further injected into partisan politics.

The Attorney General and the Department of Justice

The apparent transference in the last 30 years of partisan leadership within Administrations from the Postmaster General to the Attorney General may have been a consequence of the declining influence on patronage of the former and the continuing influence of the latter. It is time for a housecleaning in the Department of Justice comparable to that performed in the "blue ribbon" career program 20 years ago in the Bureau of Internal Revenue. Politicization of governmental legal posts cannot fail to arouse public suspicion about the even-handedness of its government; Watergate in many different ways provided substance for such fears.

The Panel recommends that every possible step be taken to remove the administration of justice and the Department of Justice from partisan politics and from the appearance of partisan politics.

More specifically, it recommends that:

· *The President appoint in the top posts of Justice only persons with high legal qualifications, and that the Senate confirm such appointments only after a thorough review of their legal qualifications;*
· *All personnel of the Department of Justice from top to bottom be covered under the Hatch Act, with its restraints on participation in partisan politics;*
· *All U.S. Attorneys be appointed by the Attorney General without limitation as to length of term;*
· *All U.S. Attorneys and their legal staffs be made part of the career service already effectively established within the headquarters of the Department, or in a governmentwide legal career service;*
· *All U.S. Marshals be placed in the civil service system.*

THE ATTORNEY GENERAL AND THE PRESIDENT

As the nation's chief lawyer, the Attorney General should not be required to represent conflicting interests. He and his department should always be in the position to render objective and reasoned legal advice to the President with respect to the latter's Constitutional duties. An Attorney General cannot do this, or at least cannot give the appearance of doing so, when acting as counsel to the President in his capacity as leader of his party or as an individual. When an Attorney General has been campaign manager for a political party, it is highly questionable whether he can readily surrender that function entirely and dissociate party affairs from the Justice Department, or keep partisan considerations from coloring his statutory functions as Attorney General.

Party legal matters would appear to be in the province of the party's general counsel, and the private legal affairs of the President the business of the President's private attorney. To the extent that there is a gray area between official legal matters and party or personal matters, and where there is a need for counsel paid by the United States Government, it is clearly more appropriate that the White House Counsel to the President, and not the Attorney General, undertake such advice.

The Panel believes that the Attorney General should be the chief legal officer for the United States Government and should be prohibited by statute from advising the President in his political or individual capacities.

THE APPOINTMENT OF FEDERAL JUSTICES

Giving the chief federal prosecutor the power to screen, recommend, and then defend the appointment of federal judges, from district courts to Supreme Court, appears increasingly questionable. The Department of Justice is probably the most frequent litigant in the federal courts. Judges should not have the impression that their performance on the bench is going to be judged by one of the litigants in cases they must rule upon. Necessarily, a sitting judge is going to be somewhat hesitant to rule as freely against a litigant who may later be called on to judge his qualifications for promotion.

The Department of Justice should, by statute, be removed from the process whereby federal judges are selected. The Panel further recommends that serious consideration be given to a substitute procedure (comparable to the so-called "Missouri Plan") whereby the preparation of initial lists of candidates would be entrusted to a bipartisan group of distinguished citizens—partly lawyers and partly laymen.

The Investigation and Prosecution of Alleged Wrongdoing in the Executive Branch

The political ties between leaders of the Department of Justice and the White House and the Department's party identification contributed to the lack of public confidence in its investigation of Watergate. This very lack of confidence occasioned the creation of the office of Special Prosecutor to investigate the Watergate affair. The ability of the President to effect the dismissal of the first Special Prosecutor underscores the need, at least at this time, to institutionalize the office.

An office of Permanent Special Prosecutor in the Department of Justice should be set up by statute and given jurisdiction to supervise investigations and prosecute all wrongdoing where an officer of the Executive, Legislative, or Judicial Branch of government is involved. The office should have authority to investigate election fraud and assist grand juries in handing down indictments whenever appropriate. Appointment to the office should be made on a nonpartisan basis, subject to Senate confirmation, for a fixed term of at least six years.

This recommendation does not go as far as the proposed legislation before the Senate (S.2803) which would establish the Department of Justice as an "independent establishment of the United States." The meaning of "independent establishment" is not entirely clear. But this Panel believes that the Attorney General and the Justice Department play such key roles in the Constitutional responsibilities of the President that they should not be removed from his overall direction. The Panel's recommendations, together with its other proposals to depoliticize the Department, would accomplish the objective of independence in the prosecution of Executive Branch wrongdoing—which is the critical area. The proposal for a Permanent Special Prosecutor should be regarded as a transitional arrangement, the need for which would wither as the Department moved from its present political role to one of a nonpolitical law office. In the interim, of course, the independent status of the Permanent Special Prosecutor must be assured by the President, the Attorney General, and the Congress.

31584

EX PARTE CONTACT WITH JUSTICE DEPARTMENT OFFICIALS

Ex parte contacts with the legislature are regulated by the Federal Regulation of Lobbying Act of 1946. The Act does not make lobbying illegal, but merely requires registration of the lobbyist. The Executive Branch, and in particular the Department of Justice, is also the subject of extensive lobbying. The so-called ITT affair is replete with special pleadings, both from within and without the Executive Branch, and is but one example of the practice.

The Panel believes that the Justice Department should move to enforce the existing requirements that all its employees keep written records of contacts by outside individuals seeking to influence the disposition of particular matters.

ACCESS TO DEPARTMENT REPORTS AND FILES

One of the most alarming concerns in the last decade relates to misuse of Department of Justice investigative files and reports. Nonetheless, White House staff members and even some persons outside of government apparently gained access to reports and files of various bureaus and divisions of the Department.

To insure that such information leaves the Department only for law enforcement purposes, statutory prohibitions on disclosure comparable to that for federal income tax returns should be enacted.

The Conflict of Roles in the Justice Department

Apart from the political problems discussed above, the Attorney General and his Department of Justice have been called upon to perform in inconsistent roles. In the first place, the Department controls the major investigatory and police forces of the national government. It has wide discretion as to which cases it will pursue, delay, or drop. It prosecutes cases for the national government, and in criminal cases that end in conviction the defendants are remanded to the "custody of the Attorney General." They are then sent to federal prisons, also within the Department of Justice. After imprisonment, their eligibility for parole is decided by boards within the Department of Justice. And Presidential pardons—or executive clemency, as the phrase has come to be known in Watergate—are usually a result of recommendations of the Department.

In other words, virtually the entire system of criminal justice of the United States is under the jurisdiction of a single officer, the Attorney General, and under one organization, the Department of Justice.

Whether or not these various powers of the Department of Justice have been abused, the possibility of abuse is present. The appearance of such a possibility is as important as its existence. A defendant might properly argue that the office which investigates and prosecutes him should not have influence in picking his jailer and parole board as well. To the knowledge of this Panel, such a combination of responsibilities under a single agency exists in none of our state or local jurisdictions. Neither does it exist in other nations of the free world.

There has not been a study for many years of the roles, responsibilities, and internal organization of the Department of Justice. Such a study is now long overdue. It is particularly desirable at this time in view of Watergate and of the long-delayed opportunity to look closely into the FBI following the death of J. Edgar Hoover.

The Panel recommends that the Congress initiate an intensive study of the entire Department of Justice in the direction of making it the principal law office of the United States government and, as far as possible, divorcing it from other responsibilities. Among the activities which should particularly be considered for transfer elsewhere are: managing the prison and parole system, operation of the program for local law enforcement assistance, and management of federal controls on immigration.

* * *

The principal objectives set forth in this chapter are: (1) to divorce as far as possible the operations of the Department of

Justice from partisan politics; and (2) to make the Department the law office of the United States and, for this purpose, to separate from it other responsibilities with which it has been burdened. It seems doubtful to this Panel that an "independent" legal department could satisfy the legitimate needs of any administration. The Attorney General should remain a loyal member of the cabinet, but he should be backed up by a highly competent and impartial corps of attorneys.

CHAPTER 5

The Public Service

THE AMERICAN government is, it is said and hoped, a government of laws. It is also a government by people. The effectiveness of the laws depends in large part upon the capabilities, judgment, and integrity of those people upon whom have been placed the responsibilities of carrying out the laws. And the quality of government hinges upon the systems and considerations whereby these people have been selected, assigned, and advanced.

The Career and the Noncareer Services

Except for the President and Vice President, all of nearly five million persons directly employed in the Executive Branch have been appointed and are assigned by someone above them in a hierarchy. There are a myriad of methods and systems whereby employment is managed, but for purposes of this report the public service may be considered in two broad categories: the *career* and the *noncareer*.

The vast majority of federal employees are in some form of career system, whether it be the civil service, the Foreign Service, one of the military services, the FBI, the CIA, the Public Health corps, the TVA, or any of a number of others. They share some degree of assurance that, if they perform capably and honestly, they may continue with the government and with their program for most of their working lives, regardless of the political party in power. Most of those in the middle and upper reaches of a career service are professionals in a special field needed or considered appropriate for the work of the agency to which they are assigned. They constitute what some have labeled the institutional staff of the Executive Branch.

The noncareer personnel of the Executive Branch embrace a wide variety of categories, but the one with which this report is principally concerned is that generally considered to be political. The political, noncareer appointees generally serve at the pleasure of the official who appointed them. They have no assurance—and a good many of them have no desire—to continue in office indefinitely. If the Presidency changes from one party to another, many of them expect to resign or be dismissed.

Political appointees may, for convenience, be considered in two separate categories. The first, very few in number, are assistants and advisers to high-level executives whose relationship is primarily personal, resting on high mutual trust and loyalty. The second are the top administrators in the government: heads of departments and agencies, their deputies, their assistant secretaries, policy advisers, bureau chiefs, and others. Some of these persons are appointed by the President subject to confirmation by the Senate. Many are generalist managers, charged with the overseeing of large organizations. Some are program specialists, experts, or prestigious professionals in the field of activity of their bureau or service, which is usually one or two or three levels below the top of a department or agency. Normally, but not always, a requisite for such appointment is party affiliation and general sympathy with the party's position in the field of activity of the particular organization concerned.

One of the problems made manifest in the hearings of the Senate Select Committee is that the demarcations between, and the roles of, these different kinds of political appointees have been blurred. Those in the first group, especially the personal and confidential advisers to the President, have been given or have assumed powers of, or over, the responsible executives under the President. Many of those in the second group, the departmental and agency executives, have been appointed without much attention to their managerial experience or competence.

There are built-in tensions in the American system of government—and probably in all systems of government. There is tension between the Congress and the Presidency, between the parties, between the houses of Congress, between the substantive and the appropriation committees of the Congress, and between the President and his Cabinet. Some of these were expected by the framers, and the divisions of powers in the Constitution were deliberately planned to accommodate them.

Within the Executive Branch there is another tension, hardly anticipated by the framers, but today operationally as important as any of the others: that between the career and the political public servants. This kind of tension is often quiescent, particularly in ongoing and relatively routine programs; however, it may acquire high voltage in new undertakings and in old ones which suddenly become politically volatile.

There are fundamental differences in attitudes, in aspiration, and therefore in behavior between political and career officials. The differences basically derive from differing perspectives in terms of time. The career officials collectively have a long memory of their programs and agencies and a considerable stake in the future of those programs and agencies. The political officers usually have little knowledge about the past and a different vision of the future of both program and agency.

Differences in memory of the past and aspirations for the future are augmented by a range of other differences:

1. In age—political officers are very often younger than their career subordinates.
2. In roles—political officers are more likely to perceive themselves as change makers of policy, program, organization, and even career personnel; career officers, or at least many of them, tend to resist quick changes or to slow them, preferring to adapt the organization gradually to new policy objectives of the Administration.
3. In relationships—political officers have ties with other political personnel in the Administration and sometimes with party leaders and sympathizers outside; career leaders relate more to each other, whom they have long known and will long know, to the appropriate

Congressional committee chairmen, members, and staff, and to clientele and affiliated professional groups outside.

4. In motivations—political officers often want to make their marks quickly to establish their reputations so they may move to more responsible political jobs or higher posts in the private sector; career officials are more concerned about their long-range program and maintaining the relationships noted above.

5. In education and experience—most in both categories have at least college degrees, and a considerable proportion of both have completed graduate programs; but, except for those political officers appointed for their program expertise, the fields of specialization of the two groups in both their education and their experience are often quite different.

6. In loyalty—loyalties of political officials are more likely to be personal to political leaders above them, up to the President; career officers like to perceive themselves as neutral so far as their work is concerned, and their loyalties are more likely to run to their agency and program, to their peers and career subordinates, and to their clienteles.

Tensions at the interface between career and noncareer officers at high levels are normal, even inevitable. They are usually most strained at the beginning of a new administration of a different party, but the widespread turnover of political officers following the 1972 reelection renewed and reenforced them. They are normally most virulent in agencies and programs in controversial fields, especially if the President is determined to make important changes in policy and program.

It is not suggested here that these tensions can or should be eliminated. The political officers bring new, fresh views to old bureaucracies, and they constitute perhaps the principal vehicle whereby a President and his party can bring about the changes promised in the election campaign. They are essential to a responsive democracy.

But the interface is itself a difficult and delicate business. If it is to be effective, there must be a free flow of communication in both directions, for each group has something distinctive and constructive to contribute to decisions. This requires that each understand the many differences indicated above, especially the difference in roles. It requires too that each party to the interface respect the other, work with and not against him, and accommodate to his views when they are valid.

Political Appointees

This is less of a problem for the career officers than for the political. The former have lived with the interface for a long time, under administrations of both political parties. But the political officers are, for the most part, unaccustomed to this kind of relationship when they first assume office. This is why the selection of political officers is one of the most important problems in American public administration. It is also one of the most casually handled. Watergate might not have happened had there been more care about the qualifications and the quality of certain politically appointed officers, or had there been an effective interface between them and mature career personnel, who would surely have advised against the activities associated with it.

Among many others, Watergate and accompanying developments made four things apparent. First, persons were placed in the highest, most difficult managerial positions in the nation with qualifications and experience for such positions that were exceedingly thin. One does not know how many or how prevalent these kinds of appointments were, but the parade of witnesses before the Senate Select Committee was not reassuring.

Second, almost none of the persons who appeared on the TV screen had any prior experience in any government, or indeed in the administration of any large organization. They displayed little appreciation or understanding of the special ethical responsibilities of public service. Again it is unknown to what extent they were representative of political appointees in general, either past or present, but the image they presented to the American people (including to the career personnel) could hardly have been more damaging.

Third, the demonstration, in the hearings and in many other forums, of suspicion, fear, distrust, and sometimes open contempt of the career servants by the political officers evidenced an attitude not conducive to an effective interface between political and career officers.

Fourth, there has been a tendency, which appears to have grown in recent years, to appoint political executives to preside over duly legislated programs when such appointees were clearly out of sympathy with the programs for which they became responsible, and sometimes even had a clear mandate from above in the hierarchy to "gut" those programs.

These four phenomena were certainly not original with the Nixon Administration. Short of a massive effort in social research, there is no way of measuring whether they have been more prevalent in the last few years than in preceding administrations. But never have they been so evident.

This Panel believes that the recruitment and selection of political appointees, and their subsequent reassignment and advancement, deserve at least as much care and inquiry as attend the same actions for career personnel. The standards for political appointees, though different, should be at least as rigorous as for career personnel. The Panel believes too that the interface between political and career officials should be made more constructive and tolerant through better understanding on each side of the roles and responsibilities of the other.

For these purposes, the Panel recommends that:

The number of posts in the subcategory of personal and confidential advisers to the President and to the heads of the principal departments and agencies under him be strictly limited by Congress through its authorizations and appropriations, and that persons in these posts be denied the power to exercise Constitutional or statutory powers beyond routine functions.

The major party organizations develop and maintain lists of the best qualified persons for possible political appointment.

The President maintain an assistant on personnel, with adequate staff, who would, among other duties, develop and maintain a continuing roster of the best qualified possible appointees to executive and judicial offices.

The primary authority and responsibility for political appointments be vested in the heads of departments and agencies who would work with the assistance of the Presidential staff suggested above, and whose choices would be subject to Presidential veto.

The Senate and its committees be more careful and thorough in their review of the nominated heads and subheads of agencies than they have been in the past; this would include nominees' attitudes toward the programs of those agencies.

Proposed appointees to the more specialized political posts be reviewed and approved or vetoed prior to their appointment by nonpartisan panels of experts in their fields. (See p. 75.)

New political appointees be encouraged or even required to attend educational briefing sessions concerning their responsibilities and particularly their relations with career personnel.

A further problem with respect to many political appointees stems from their generally short tenure in any given position or assignment. This is due to many different factors: administration decisions to reassign them from one job or one agency to another; the desire to leave for more rewarding or lucrative employment in the private sector; dissatisfaction with the demands, difficulty, and relative unpleasantness of the work; inability to adapt to a very different environment and set of relationships; and inadequate performance, or policy or program differences which result in their being encouraged to depart. Whatever the causes, the rapid movement in and out of key positions, in and out of different agencies, and in and out of government would be intolerable in most private enterprises.

Rapid rotation and replacement of political officials is by no means a new problem. A recent study of the tenure of undersecretaries and assistant secretaries appointed during the terms of Kennedy, Johnson, and the first Nixon term showed that about one-fifth of them served less than 12 months, and fewer than half served more than two years. The wholesale changes that followed the reelection of the President in 1972, and which still continue, must surely have further reduced the average tenure. Yet, many observers have set one or two years of experience in such posts as essential before an incumbent is prepared to pay good dividends. Former Commerce Secretary Maurice Stans, for example, said: "A business executive needs at least two years to become effective in government, to understand the intricacies of his programs, and to make beneficial changes." [1]

[1] *Business Week*, September 22, 1973, p. 12.

A secondary effect is the impact of this turmoil upon the career personnel, upon their ability to maintain some degree of operating continuity, let alone contribute to policy decisions, and upon their morale. As one career man in the OMB said: "The top management of this place is like a carousel. I don't know from week to week who my next boss will be."

The Panel recommends that every political appointee to a permanent position commit himself, at the time of appointment, to serve for at least two years.

It further urges that this and subsequent administrations slow down the pace of reassignment and rotation to the extent possible.

The Civil Service

> *I believe a strong career service is one of the greatest strengths of our democratic process, and one of the best guarantees of sound, effective and efficient government—even more so in 1974 than in 1883.*
> —Vice President Gerald R. Ford [2]

Although, to their everlasting credit, almost no career public servants were directly implicated in Watergate activities, the career services were indirectly related in at least three significant ways. First, one source of the difficulties was the failure of the interface, described earlier, in some sectors of the government, which grew out of disparagement and distrust on each side toward the other. Second, the loss of confidence of many, probably most, American citizens in their government which resulted from Watergate applies to the whole Executive Branch (and perhaps to the other branches as well). Most Americans do not make nice distinctions between political and career personnel. Third, Watergate evidently has contributed to impairment of the confidence, including self-confidence, and morale among the career personnel themselves. In sum, the problems of the interface have probably been exacerbated.

2 Remarks made during Ceremonies Commemorating the 91st Anniversary of the Federal Civil Service, January 16, 1974.

Five out of six of the nonuniformed personnel of the United States, more than two and a half million persons, are employed under the civil service system, and it is to these that primary attention is here directed.

ORIGIN AND DEVELOPMENT OF MERIT PRINCIPLES IN THE CIVIL SERVICE

The federal civil service system dates from the Pendleton Act of 1883, an Act which resulted from growing doubts about spoils and patronage, and was finally triggered by the assassination of a President by a disappointed office seeker. Its main thrust was to establish and maintain a career civil service on the basis of merit and free of political pressures and abuses. From the beginning it was premised on the assumption that persons assured of a permanent career could faithfully and loyally perform the duties of public office regardless of the party in power.

In the nine decades since its origin, the civil service system has grown persistently, though sporadically, and not without stormy periods, from a coverage of about 10 percent to about 85 percent of the federal civilian service. It has become perhaps the chief instrument for providing continuity in our pattern of government. It has imparted institutional stability and memory in adapting public management to the differing policy objectives of successive administrations. It has increasingly supplied the professional and managerial experience and expertise which are indispensable to successful program performance. With few exceptions, career civil servants understand that the government belongs to the people and that they are trustees, not proprietors, in the use of public authority. Presidents and their political aides increasingly learned that their own managerial effectiveness depends in large part on the support and the effective use of the corps of career personnel.

THE EROSION OF MERIT PRINCIPLES

All recent Presidents and their high political aides have expressed impatience and even exasperation with career civil servants on grounds that they were resistant to change, especially in programs that a new administration urgently wanted to get done

—or undone. Many political officials felt to some extent that numerous aspects of the civil service laws, regulations, policies, and practices were products of an age of simpler government, and impediments to the achievement of their broad policy goals. One device to insulate agencies from the strictures of the civil service system was to establish new agencies, totally exempt from the system (though many were later blanketed in). This was massively used by Roosevelt in his many alphabetic New Deal agencies, and was more recently used in the creation of agencies in new and different fields of activity, such as the Atomic Energy Commission. Most of these exemptions from civil service were endorsed, if not insisted upon, by the Congress. Another technique was the exemption of a substantial number of key positions from the normal competitive processes, a practice formalized under Eisenhower in an exempt category then and still known as Schedule C. A third device was, and is, to superimpose above the career officers a new layer of political appointees sympathetic with the incumbent administration and empowered to overrule their career subordinates.

Finally, there are a variety of devices to intervene in and manipulate the management of the career civil service itself for partisan ideological purposes. These include: (1) the requirement of political clearance of new career appointments, promotions, assignments, and other personnel actions by the White House, by agents of the White House in the agencies, and by national, state, and local party organizations and politicians; (2) misuse of the technique known as selective certification to enable the appointments of "loyal" persons who are not near the top of competitive registers; (3) the use of classification authority to arrange preferences in promotions and also for downgrading of positions to demote—or encourage the resignation of—persons considered disloyal in the personal or ideological sense; and (4) the use of reorganizations to eliminate the "disloyal" and make places for the "loyal."

Such practices as those described in the preceding paragraph are most serious when applied to the higher civil service positions. They undermine the merit principles upon which the career civil service is based; they subvert the confidence and morale of the career personnel who remain in the service; and they add to the crisis of confidence of the American people in their government. Most of them are clearly illegal.

None of these devices is new. Short of a thoroughgoing, nonpartisan survey, it would be impossible to establish that their use is substantially greater now than in previous administrations. But there is evidence that they have significantly changed, and are changing, the "tone" and the morale of the public service. For example:

· The politicization of many of the regional officials of the various domestic departments and agencies, most of whom had previously been experienced career personnel;
· The insertion of a layer of political appointees in the OMB, heretofore a model of experienced, even-handed competence;
· The wholesale replacement of all of the Assistant Secretaries for Administration, most of whom for 20 years had been considered nonpartisan career officers, with politically endorsed new appointees, and the politicization of many of the specialists under them in staff fields such as personnel, finance, budgeting, training, program evaluation, and inspection;
· Questionable personnel actions in several agencies, as disclosed by recent investigations and actions by the U.S. Civil Service Commission.

The need and importance of a modest number of top political officers have already been emphasized. This proposition does not justify conversion of a large number of hitherto senior career positions and subordinate positions far down the civil service scale for partisan purposes. Neither does it support the imposition of political tests for the upper levels of the civil service, a practice by no means new or unique to the present Administration. The critical error of the present Administration apparently has been to assume that, since large numbers of the senior career officers were appointed during regimes of the other political party, old loyalties were so fixed that they could not be transposed to the objectives of a new and different administration. This was a miscalculation with serious consequences.

The Panel recommends that the Committee urge the Congress, the President, and the U.S. Civil Service Commission to require and superintend strict enforcement of the laws and regulations forbidding political considerations in career personnel actions.

A NEW PERSONNEL SYSTEM FOR SENIOR CAREER AND POLITICAL OFFICERS

To no small degree, the invasion of politics into the senior ranks of the civil service has its origin in understandable frustrations about the procedural requirements of the present system. There is no doubt that it limits the discretion of the President, agency heads, and their principal political associates in the responsive management of their programs. Authorization of a new personnel system for administering senior career positions and their political counterparts would correct some undesirable and some illegal practices. The idea is not new, and to its credit, the present Administration has proposed such action.

Professional public administrators, including the top ranks of the civil service and students of administration, have long advocated a public personnel policy which provides the flexibility to insure that the top executives and professionals in the federal government are properly responsive to public policy as enunciated by the President and the Congress. The desirability of granting broad discretion to policy leadership in the choice of key subordinates is not in dispute. It is an essential principle of modern manpower and program management.

The existing civil service rules governing senior career positions foster unacceptable rigidity. It is difficult to give full weight to the impact of the man on his duties when his salary level must follow a statutory series of grades and steps within grades, and his job must be classified at a given grade. Inflexibility in the use of senior individuals is also hard to overcome under a system which strictly limits the total number of so-called supergrade positions in grades GS-16, 17, and 18. The authorized number is believed by many to be inadequate. Agencies are therefore reluctant to give up any supergrade slot to another agency, no matter the priority of need.

In lieu of an unsatisfactory, no longer adequate Schedule C, an

74

inadequate number of top civil service positions, and an inflexible system for administering both, the Civil Service Commission, with the President's approval, proposed legislation for a new senior personnel system known as the Federal Executive Service. In briefest summary, it would: (1) establish a statutory base for career and noncareer officers at the senior program and staff levels; (2) provide an independent review of the qualifications of career officers by boards of specialists established for the purpose; (3) provide managerial authority to assign and reassign them for their most effective utilization; (4) provide a flexible basis for determining their pay, assignment, and status; and (5) give protection to senior career officers.

The Panel supports in principle the basic features of the Administration's proposed Federal Executive Service. The Panel further recommends Qualification Boards for specialized political as well as career appointments.

THE CONFLICTING ROLES OF THE
U.S. CIVIL SERVICE COMMISSION

A great variety of laws and executive delegations have imposed upon the U.S. Civil Service Commission an aggregation of responsibilities of different and inconsistent kinds. The result is a conflict of roles in a single agency, in some ways comparable to that discussed earlier in the Department of Justice. Its duties are a strange mixture of quasi-judicial, quasi-legislative, and administrative functions. On one hand it is a Presidential staff agency, guided and constrained by the policies of the incumbent President. It is also a principal instrument in establishing or proposing legislation and regulations applying to the civil service. And it is the major enforcement and appellate agency for assuring compliance with those laws and regulations.

The Panel is not prepared to make specific recommendations about the reorganization of civil service administration, but it does earnestly urge that, in the personnel study proposed below, serious consideration be given to the establishment of a new and separate agency for the monitoring, investigative, and adjudicatory functions such as those involved in: (1) the maintenance of merit standards; (2) the prevention of political influence upon competitive civil service positions; (3) serving as an appellate court on disciplinary actions, allegations of discrimination or unfair treatment, and disputes on classification; and (4) serving as an ombudsman for federal employees on matters of personnel administration generally.

A STUDY OF THE CIVIL SERVICE

There has not, in a very long time, been a thorough study of the federal personnel systems, and particularly of that largest part of them which comprehends the federal civil service. In the interim a multitude of ad hoc laws, regulations, rules, and practices have been adopted which are so complex as to defy comprehension, let alone understanding, by more than a very few. There have been studies of certain aspects of the personnel system, such as the recent one on classification and pay practices. But the most recent overall survey was that of the Second Hoover Commission, almost 20 years ago. There can be no group upon which the welfare of American citizens depends to a greater extent than the federal civil service—for its integrity and its effectiveness.

The Panel recommends that the appropriate bodies in the Congress conduct or advocate and support a thorough and comprehensive study of the federal civil service and consider and act upon the recommendations which it produces. The study should extend, among other matters, to the existing laws and rules governing political incursions on the merit system, many of which are outdated, and to the central organization for the management and protection of the career civil service.

* * *

America has had more probity, ability, and commitment to the public good by the vast majority of its civil servants than recent disclosures might suggest. Intrigue and peccadilloes are not the distinguishing characteristics of civil servants. Neither are willful inefficiency, stubborn opposition to change, and intentional attempts to undermine political superiors. Over 90 years of experience have produced incontrovertible evidence that the career services will discharge assigned duties responsively and capably when they are responsibly and capably led.

CHAPTER 6

Information, Disclosure, Secrecy, and Executive Privilege

AMONG the most troublesome of the issues confronting the United States today are those concerning access to information about executive processes. As a general proposition, a society committed to liberal democracy cannot adopt or defend any position other than one of maximum feasible freedom of information. And yet this freedom, like other freedoms, cannot be absolute in a modern and civilized society.

In the practice of government in the United States, we have come to recognize several competing needs and actions. Insofar as freedom of information is concerned, these competing needs involve national security, secrecy, and executive privilege on the one hand, and openness in the conduct of public affairs on the other.

Long ago in the *Nicomachean Ethics,* Aristotle argued that in any consideration of virtue there was a danger of extremes. An extreme of virtue might turn out to be a vice in the relationship of one man to another, or of one man to a community. The challenge to the virtuous life was to find the happy balance between an extreme of virtue and extreme of vice. Moral virtue, Aristotle held, was a mean between two vices: the vice of an excess of moral fervor which would tolerate no difference, and the vice of deficiency of moral fervor which would acknowledge no standards.

It is within this context of the need for balance that the Panel has sought to isolate the major issues arising in the wake of Watergate and to comment thereon. Clearly, within the existing limits of time and space, we cannot hope to deal with this complex subject in any great detail or specificity.

Secrecy in Government

Trends in the conduct of government culminating in Watergate point up the need for rebalancing the conflicting requirements of openness in the operation of the public business and secrecy in governmental decision-making processes. It is imperative that the integrity of those processes be protected and that the right to privacy of individuals not be abridged.

In a recent unpublished paper, Harlan Cleveland and Stuart Brown make a number of important points about secrecy in government—particularly in foreign affairs. One of their major theses is that the withholding from the public of information by public officials has a tendency to corrupt government processes:

> Once the system permits the President and his agents to decide who should know what about Executive intelligence and operations, it is overwhelmingly likely that Government officials will hide their mistakes and their debatable judgments from colleagues, subordinates, inspectors, controllers, Congressmen, courts, and constituents by deciding that none of those have a "need to know." [1]

The Panel agrees with this conclusion and further believes that the escalation toward more and greater classification of information since the onset of the country's international difficulties following World War II has spilled over into purely domestic affairs. In spite of a series of Executive Orders issued over the past 15 years which have sought to restrict the authority to class-

[1] Harlan Cleveland and Stuart Brown, "The Limits of Obsession" (Paper prepared for the conference sponsored jointly by the Senate Select Committee on Presidential Campaign Activities and the Center for the Study of Democratic Institutions, Santa Barbara, California, December 3, 1973), p. 11.

ify documents and to provide for their declassification as appropriate,[2] the trend toward secrecy in both foreign and domestic matters appears to be unabated.

Secrecy presents problems in all branches of government. It is natural for anyone to prefer to publish, even advertise, what shows off to advantage or strikes a popular note, and to suppress or play down what is to the contrary. How far such preferences prevail depends, among other things, upon the practical maintenance of a monopoly control of the information in question. The framers of our Constitution achieved that monopoly in 1787 and produced their own final document after a summer's deliberations in entire secrecy by the simple expedients of agreeing not to talk outside of the Convention until their work was done, and then holding to that resolution.

The immense mass of information gathered and generated in the Executive Branch, and particular items of news within it, are not so easily controlled. Conscience, vanity, greed, and other human frailties among the gatherers and generators may impel disclosures. Superiors, subordinates, and counterparts or rivals in other agencies may judge matters differently. Members and committees of Congress and their staffs, newsmen, and lobbyists thrive on discoveries. The generator has the tactical advantage of the initiative. When he decides on silence, or stamps his document "secret," the trick is to reinforce or frustrate his initial monopoly according to judgments of the balance of public interest, and to institutionalize processes and safeguards to that end. Standards and criteria are needed, but general propositions do not decide concrete cases.

The Panel believes that the series of Executive Orders issued over the past 15 years that have purported to restrict the authority to classify documents and provide for their declassification are insufficient to protect the public interest in disclosure. Congressional review and strengthening of these Orders are in order.

The Panel is certainly not disposed to recommend any dramatic overturn in the practice of protecting from public view matters which would compromise the welfare of the nation or the rights of

2 As recently as E. O. 11652, March 10, 1972.

its citizenry to privacy. But public officials conducting official business privately need to ask themselves continually whether a curtain shielding their transactions is really necessary or merely convenient, and, how they would look if the curtain were involuntarily lifted.

In short, the Panel feels that the trend over many years has been to withhold information from public view unnecessarily; that there continues to be a need for some degree of secrecy in government affairs; and that a redress in the balance is a continuing responsibility of all three branches.

National Security

One of the questions prompted by post-Watergate events is whether it is possible to define national security with sufficient precision that it cannot be used as a blanket to cover illegal acts and/or usurpation of power by public servants. William Watts describes the indiscriminate use of "national security" as the "ultimate fig leaf." [3] Most observers would agree that in several instances, such as the illegal activities of the "plumbers" and the bombing in Cambodia, the concept has been grossly misused to cover executive actions which either should not have been undertaken in the first instance, or, if properly undertaken, should have been open to the scrutiny of public debate.

The definitional question has concerned a number of writers in the field. For example, Klaus Knorr has written that "national security" is "an abbreviation of 'National Military Security'—to denote a field of study concerned primarily with the generation of national military power and its employment in interstate relationships." He would specifically exclude "domestic security" from the field. [4]

[3] William Watts, "The Limits of National Security" (Paper prepared for the conference sponsored jointly by the Senate Select Committee on Presidential Campaign Activities and the Center for the Study of Democratic Institutions, Santa Barbara, California, December 3, 1973).

[4] Frank N. Trager and Phillip Kronenberg, *National Security and American Society* (Lawrence: University Press of Kansas, 1973), p. 6.

Information, Disclosure, Secrecy, and Executive Privilege

Others, notably Carl Figliola in his in-depth review of the current national security literature, concludes that most authorities would favor a definition which "incorporates both international and domestic considerations." [5]

Frank Trager and Frank Simonie, for example, offer this definition: "National Security is that part of government policy having as its objective the creation of national and international political consideration favorable to the protection or extension of vital national values against existing and potential adversaries." [6]

The Panel considers that there are domestic issues which are sufficiently delicate in nature and sufficiently intertwined with our international affairs to warrant the cover of secrecy implicit in a "national security" classification, and therefore properly protected from public disclosure by the executive. It concludes, therefore, that a definitional exclusion of internal affairs is not an adequate answer for protecting the system against the misuse of the power to cover illegal or improper domestic activities.

Rather, the Panel concludes that continued vigilance on the part of executive officials, the Congress, the media, and the citizenry, combined with appropriate judicial review, is the only answer we presently have to deal with executive misuse of "national security."

Executive Privilege

Another question precipitated by post-Watergate events concerns the proper limits to the exercise of the executive privilege, and how these limits can be set and enforced. Few would contend either that the right of the President to invoke executive privilege is absolute or that no such right exists. Although the term itself is of fairly recent origin, the concept that the Executive may, under certain circumstances, withhold information is deeply embedded in our tradition and the administrative realities of life. The Panel

[5] Carl L. Figliola, "Considerations of National Security Administration: The Presidency, Policy Making and the Military," *Public Administration Review* 34, no. 1 (January/February 1974): 82–88.

[6] From an article in Trager and Kronenberg, *National Security*, p. 36.

is clearly of the view that, on the negative side, the concept does not—and should not—extend to the concealment of illegal acts by the President or his subordinates. On the positive side, we are convinced that the President must be able to consult with advisers of his own choosing on official matters without the necessity of public disclosure of those conversations.

The Panel agrees that, in such an impasse as we have recently seen involving executive privilege, the courts represent the proper institution for resolution of the issue. In a recent U.S. District Court ruling, Judge Charles R. Richey said: "Any evidence which concerns the government's illegal acts is not privileged—the claim of executive privilege is not absolute."

The Panel supports this view and concludes that, except in the case of impeachment, where contests about executive privilege cannot be resolved by the President and the Congress, the Judicial Branch represents the proper arbiter of the issue if it is brought before the courts and they accept jurisdiction.

At the same time, the Panel hopes that the courts and the Congress will not lightly seek access to Presidential documents. Our tradition and the normal comity among the branches have long protected the nation from the trauma inherent in the conflict over this issue. We need a speedy return to relationships which once served us well.

Freedom of Information

It has become increasingly apparent that the Freedom of Information Act, Public Law 89–487, requires adjustments to make it more responsive to the needs of our society. The Act imposes four primary obligations upon all executive agencies of the federal government. The first obligation is to publish in the Federal Register descriptions of organizations, statements of procedure, rules of procedure, substantive rules of general applicability, and statements of interpretations of general policy.

The second obligation upon all agencies is to make available for public inspection final opinions and orders arising in the adjudication of cases, statements of policy, and interpretation not published in the Federal Register and staff manuals and staff instructions affecting the public.

In the third place, an agency has the obligation to maintain and to make available to the public a current index identifying information available to the public for any matter issued, adopted, or promulgated after July 4, 1967.

Finally, an agency has the obligation to make its records promptly available to any person. Such availability may be enforced through a court of law.

The Freedom of Information Act does contain a series of exceptions to its requirements: (1) matters required by the Executive Order to be kept secret in the interest of the national defense or foreign policy; (2) matters related solely to internal personnel rules and practices; (3) trade secrets and commercial or financial information obtained from a person and considered to be privileged or confidential; (4) personnel and medical files, the disclosure of which would be an invasion of privacy; (5) investigative files compiled for law-enforcement purposes; (6) reports of an agency responsible for the regulation or supervision of financial institutions; (7) geological and geophysical data; and (8) interagency and intra-agency memoranda and letters not normally available by law to a party not engaged in litigation with an agency.

The general intent of this Act of 1966 is clear. It was to provide maximum feasible access by all citizens to information about the policies, organization, procedures, and actions of the agencies of the United States Government. The objective is that the affairs of government be conducted openly and in the light of full public disclosure. Unfortunately, a law declaring the principle of freedom of information is not enough. Administrators, the chief executive, legislators, and others must accept and support the obligations of the law, in letter and in spirit.

The conduct of the federal government by the Executive Branch in recent years has tended to deny and to abrogate the legal obligations of the Freedom of Information Act. This is one of the principal issues raised by the Watergate hearings.

There is undoubtedly a need at the present time for two kinds of action by Congress: (1) to reaffirm the principle of the Freedom of Information Act; and (2) to review the exceptions provided in the Act and to restrict their applicability. In particular, the language concerning exceptions to the obligations of publishing, making available for copying, indexing, and responding to requests for information needs to be rewritten and clarified. The exceptions are too broadly worded and not clear in meaning, especially the exceptions related to "agency memoranda and letters" and to data exempted because of their relationship to "national defense and foreign policy."

Unauthorized Disclosures

One of the most complex issues to be resolved in the aftermath of Watergate concerns the resort by public officials to unauthorized disclosures of information which has been restricted by appropriate authority. Given the imperfect state of laws, organizations, and men, it appears that the role served by such disclosures must be recognized. The nature of bureaucracy, whether public or private, is such that the penchant for secrecy (or the disinclination to share information) is an everyday fact of organizational life. Information represents power.[7] Bureaucracies and their constituents often develop a system of "preferential" sharing of information to the mutual benefit of both parties, sometimes to the detriment of the general public. Thus, railroad executives have easier access to the Interstate Commerce Commission, as do Farm Bureau officials to the Department of Agriculture, than does the ordinary citizen. This system is reflected in the frequently close affinity between bureaus and agencies and the chairmen of Congressional committees or subcommittees having jurisdiction. Information passed in these channels, though not prohibited by law from disclosure, often is not available to the public or even to other members of Congress.

Even newsmen, those guardians of the "public's right to know," tend to support the operation of this informal system of preferen-

---

[7] Francis E. Rourke, "Bureaucratic Secrecy and Its Constituents," *The Bureaucrat* 1, no. 2 (Summer 1972): 117.

tial disclosure while publicly making noises of righteous indignation:

> . . . there are significant ways in which newsmen . . . benefit from executive secrecy and have on occasion been very supportive of it. Such secrecy gives some reporters, particularly columnists, an opportunity to publish information obtained from inside sources that would be of little value if it were freely available to the public at large.[8]

This raises the question of who restricts disclosure, where such sharing is not clearly covered by statute. Why is it legitimate for the secretary of a department to leak information to a senator favorable to the department's (or the secretary's) position, and not legitimate for a less august public official to do so *when he sincerely believes that a serious wrong is being done to the public interest?* The practical answer is straightforward: organizations cannot function if the members are not team players and if all exercise their individual, independent judgment irrespective of its relationship to agency policy and direction. Though true, this approach to the question is too simple. Each individual issue or instance of unauthorized disclosure involves both a moral and a practical choice.

In spite of the existence of the Freedom of Information Act, which attempts to open up nearly all government procedure and actions to the public, a large part of the decision on what can be disclosed is an administrative one made within the hierarchical structure of the agency. Officers and employees of the United States are still proscribed by law from disclosing confidential information, except as "provided by law." [9] The term "unauthorized disclosure" covers everything from a Daniel Ellsberg, who was accused of being in violation of the espionage laws, to an A. Ernest Fitzgerald, who gave information embarrassing to his agency (the Air Force) in testimony before a Congressional committee. Such acts of conscientious disclosure, as contrasted with the providing of information for personal financial gain, have been dubbed "whistle blowing." [10]

---

[8] Ibid.: 119.

[9] 18 U.S.C. 1905.

[10] For a series of cases of "whistle blowing," see Charles Peters and Taylor Branch, eds., *Blowing the Whistle: Dissent in the Public Interest* (New York: Frederick A. Praeger, Inc., 1972).

WATERGATE

... whistle-blowing boils down to an attempt to suspend the rules that produce loyalty and cohesive behavior [in] institutions. . . . When a whistle-blower suspends the rules, he is arguing that extraordinary circumstances exist to justify what he is doing. In short, he throws the case open to debate from scratch—debating politics, objectives, and first principles—because he feels that the rules and guidelines for resolving disputes and failures within an organization have been insufficient.[11]

Unfortunately, there is no neat, blanket solution to the dilemma of unauthorized disclosures. Given the conflict between the intent of the Freedom of Information Act and the normal bureaucratic tendency to promote secrecy, there will continue to be a need for "whistle blowers" who call attention to alleged illegal acts, and whose actions can be judged on how well they serve the public interest only after the fact and on a case-by-case basis. There is no real alternative to placing trust in public servants of intelligence and integrity who have a deep commitment to serve the American public, whether they be career employees, appointed officers, or elected officials.

Each case helps establish that citizenship requires a difficult balancing of many loyalties, and that people must take a personal responsibility for the larger context of what they do and are involved in. . . .[12]

[11] Ibid., pp. 290–291.
[12] Ibid., p. 297.

CHAPTER 7

The Financing of Federal Political Campaigns

FEW CITIZENS of the United States would disagree with the assertion that among the most disturbing revelations of Watergate have been those related to the financing and conduct of our federal political campaigns. An overwhelming majority favors action to prevent recurrence of the scandalous features that infect the system by which money is raised and the uses to which it has been put.

But essential reforms will be hard to obtain. In the first place, modern election campaigns are necessarily costly. Further, there are two other impediments to change: (1) the subject of election finance in our federal system is complicated, lending itself to no simple solutions; and (2) powerful vested interests, both private and public, oppose corrective legislation likely to have significant effect.

Scandals involving campaign finance are not new in American politics, but the dimensions of such phenomena in recent months far exceed all precedents. The past year saw the first occasion when a Vice President of the United States has resigned, pleading *nolo contendere* to a charge of felonious tax evasion on income derived in part from funds allegedly obtained for campaign purposes. At this writing, eight major corporations have accepted

WATERGATE

guilty pleas and been fined $5,000 each for criminal violation of
the prohibition of corporate campaign contributions in the federal
Corrupt Practices Act. Seven top corporate executives have been
fined $1,000 each in these cases. The former head of a large labor
union has been sentenced to a three-year prison term for a similar
offense. And the conduct of numerous other corporations and
unions in the 1972 campaign is under intensive investigation.
Convictions for such offenses are rare in our history.

Rapid escalation of campaign costs has placed severe pressures
upon candidates and party organizations to obtain larger sums of
money than ever before—even by dubious means and from questionable sources. Funds collected and monies spent have reached
new high levels in every federal election year, and in state and
local elections. Charges of extortionate practices in campaign fund
solicitations, and charges that governmental actions—both executive and legislative—are heavily influenced by campaign finance,
circulate through all the news media and are commonplace in
private conversations. When such charges are proved, or even
widely believed, confidence in the quality and integrity of government is lost.

There is a wholly legitimate objective in campaign funding.
Indeed, financial resources are essential to the proper functioning
of political processes in this country. A democratic republic with
universal suffrage can hardly hope to succeed without a well-
informed electorate, familiar with key personalities and the increasingly complex policy issues present at all levels of government—national, state, and local. To bring contested policies and
contrasting candidacies before the electorate is an obvious necessity; it is equally obvious that to do so must involve considerable
expense.

That observation leads to the primary issue: How can the funds
essential to inform the entire electorate be raised and spent without recourse to past abuses and without victors heavily beholden
to any special interest?

This and related problems come to the fore now because of
revelations concerning Watergate and a host of related misdeeds.
The concepts involved apply also to state and local elections, but

88

reference herein will be made to these only in cases where their conduct is intertwined with federal elections.

Legitimate objects of expenditure can be classified under three heads: (1) construction of an accurate list of eligible voters—the registration process; (2) the mechanics concerning elections (and primaries) as to times, locations, ballots or voting machines, counting votes, reporting results, and so on, designed to assure accessibility, the secret ballot, and integrity in the electoral process; and (3) all those means used to acquaint the electorate with the policy issues at stake and the individual candidacies involved.

It would seem beyond dispute that whatever sums are required to ensure the propriety and integrity of registration, primary, and election mechanics can be readily justified. The cost of these activities, although significant, is dwarfed by the funds expended on campaigning per se.

Some $400 million was spent in 1972 from the campaign funds of political parties and candidates, not including spontaneous and unauthorized expenditures by interested individuals and organizations, acting outside the standard framework. About one-third of the $400 million was spent on Presidential primaries and elections, and a large share of the remainder went to contests for the Senate and the House of Representatives.

As to the levels of campaign financing, two facts have a direct bearing. One is that amounts spent doubled in eight years, from 1964 to 1972—an escalation not fully explainable in terms of higher costs for television, postage stamps, population growth, and so on. The other is that overhead costs of solicitation and collection consumed a considerable part of the $400 million, through direct mail appeals for money, television appeals, testimonial dinners, bull roasts, and clambakes. Solicitation costs consumed about half of the $30 million spent in the final 1972 campaign by Mr. McGovern, and perhaps one-fourth or more of Mr. Nixon's $60 million funds. Public solicitations are a form of campaigning, but sums remaining after deduction of such costs provide comparatively little for presentation of platform positions and personal qualifications to the electorate. There are obvious ways by which the burden could be held in check, in addition to cutting back solicitation costs. The brevity of English parliamentary campaigns, for example, helps to make them less expensive. Yet the size of the grand total spent is less important than the abuses that have grown up in collection and use of these funds.

Campaign Expenditures That Are Proper Public Charges

Expenditures involving the mechanics of elections are certainly for an essential public purpose. In some cases local governments, acting under the provisions of state law, skimp on basic requirements. Voting machines are not in universal use, and protections against corruption or intrusions upon secrecy are not always adequate. It may be hoped that the states will correct these deficiencies, making further federal action unnecessary.

Preparation and distribution of complete and accurate registration lists are also state-local functions, but performance is so deficient that political parties and other concerned organizations expend much energy and some money to register eligible voters who would otherwise be unable to vote.

The cost of a universal registration system is properly a public charge which, if the obligation is not fully met at the state-local level, should be assumed forthwith by the national government— at least for all national elections.

Another election function, not now widely used, is clearly worthy of public funding. The State of Oregon and some other jurisdictions pay for campaign brochures—mailed to all voters —containing the platform positions and arguments advanced for each candidate. The materials are prepared by the contending parties, with equal space allowed to each.

Such brochures should be circulated, at federal expense, at least two weeks in advance of every primary and final election for federal office.

Responsibilities of the News Media

The media contribute heavily to popular understanding of public affairs, through their normal functions of factual reporting, editorial commentary, and occasional analyses in depth. They are accused, variously, of excessive care in avoidance of offending business firms that purchase advertisements and commercials; of an imbalance in news coverage or editorial policy because ownership is dominantly affiliated with one party; of lack of courage in facing threats or intimidating gestures; and of a critical bias against all public office holders—especially toward elective officials. Yet their day-to-day reporting of events before, during, and after campaigns renders an invaluable service to the nation.

Nevertheless, a larger contribution could be made by the media, without placing any burden upon campaign funds. Some newspapers publish "battle pages," which are presentations of statements made by opposing candidates (or parties), side by side in common space, during the course of a campaign. The Kennedy-Nixon debates of 1960, broadcast as a public service, are now legendary, but the "fairness doctrine" has since been interposed to prevent subsequent use of this format, because numerous independent and minor party candidacies make it unwieldy and unworkable.

The legislative barrier in Section 315 of the Communications Act against TV or radio debates between main contenders should be lifted, freeing commercial broadcasting to exercise discretion in equitable presentation of candidates and issues.

More recently, an important additional political forum has emerged. The educational programs of the public broadcasting system appear to be highly useful, but freedom of expression is now imperiled by the threat that federal financial support may be curtailed or withdrawn.

Generous public funding of public broadcasting for political educational purposes is fully justified, subject to reasonable rules of equitable treatment.

If expanded use of broadcast channels becomes available without charge, pressures upon campaign finance would be correspondingly lessened. Beyond doubt, the emergence of television has revolutionized American politics. Although use of this medium of communication at commercial rates involves great and increasing expense, access to it is considered imperative by all candidates, and most of all by newcomers to the political scene. Recent legislation strictly limits amounts spent on broadcasting in federal elections to $.06 per person of voting age within the constituency. This will help to prevent excessive competitive spending.

Governmental Support for Other Essential Campaign Activities

The strongest argument for direct federal appropriations to finance campaigns for federal office (in whole or in part) is negative. That is, it rests on the proposition that present practices are intolerable and that alternative means must be found to replace them. From President Theodore Roosevelt onward, the alternative of public financing has been advocated by many of those seeking to remove improper influences affecting both Legislative and Executive Branches of the federal government.

The case for expenditure of governmental funds in connection with campaign finance, as such, is enhanced by the fact that such support is already provided indirectly through the recently enacted tax-credit and tax-deductibility features of the federal income tax system. This change recognizes the propriety of federal encouragement of political contributions, at least in modest amounts. Several states have already followed this lead in their own tax codes.

Arguments against federal financial support are chiefly these: (1) such expenditures would place an improper and unnecessary burden upon taxpayers; (2) they would discourage citizen participation and weaken party morale; and (3) constitutionally, the choice of Presidential electors is solely a state responsibility, as are Congressional election arrangements, so that federal intervention in these matters is unconstitutional and unjustifiable. (The minor amount of public support here contemplated pales into insignificance beside an annual budget of $300 billion.) Neither the case against propriety nor that against constitutionality is convincing, however, and the argument concerning loss of citizen participation carries little weight against a system of mixed public and private campaign finance.

It is the unanimous conviction of the Panel that federal appro-

priations in support of campaign funding for national elections would serve a proper and desirable public purpose.

Such support is essential to restore some sense of equilibrium in national politics and to restore public confidence in the integrity of the electoral process. It should certainly extend to Presidential primaries as well as to final Presidential elections.

Equally, it should extend to final elections of Senators and Representatives. The need for federal support for Congressional primaries is not as obvious, particularly since party nominees may be chosen by conventions—a far less expensive procedure. There appears to be no urgent need of federal support for conventions, as such, at any level.

GOVERNMENT SUPPORT:
EXCLUSIVE OR AUXILIARY, AND IN WHAT AMOUNTS?

The national experience with exclusively private campaign financing provides nearly conclusive evidence against the method. That does not establish a case for exclusive public financing, however. There may, indeed, be a constitutional right for individual citizens to contribute toward campaign funds or spend money independently in support of causes and candidacies—at least within reasonable limits. Active and widespread citizen participation in the electoral process is clearly desirable, and campaign giving is one form of participation. This reinforces the case for a mixed system of campaign finance—part public and part private.

Assurance of a minimum level of public funding would relieve candidates (and parties) of the sense of desperation enhancing temptation and dependency upon powerful special interests.

The value of direct public support to the candidate is much greater than appears on the surface. Ten thousand dollars (or, in a Presidential contest, a million dollars) automatically available immediately upon nomination is worth two or three times as much as it would be late in the campaign, because it can be carefully and prudently scheduled to gain the maximum benefits. Incumbents, as well as challengers, would feel less anxiety that every action or position taken must be weighed against gains or losses in campaign funds. Most of all, the candidate's time and energy can be turned, to a very considerable extent, from fund raising to serious public issues.

Public law 92–178 provides a level of support of $.15 for each potential voter for each major party Presidential candidate during the course of the final election campaign. This would produce between $20 million and $25 million for each major party candidate, or roughly half the *average* 1972 expenditures.

The Revenue Act of 1971, as amended, provides opportunity for each federal income tax-payer to authorize use of $1.00 of his or her payment for support of presidential campaign funds. This "check-off" arrangement produced small response on tax returns for 1972, but early reports indicate a strong response on 1973 returns, leading to an expectation that total yields would be fully adequate to cover the maximums set in the Act ($.15 for each potential voter for each major candidate). Recent estimates indicate that the income tax check-off system will generate substantial amounts by 1976. However, this money will be subject to the regular appropriations process (including a possible presidential veto) and recipients could not raise private funds.

The Senate has voted in favor of "matching funds" for those Presidential primary candidates who have raised at least $100,000 from small donations. Further, the Senate has voted for federal appropriations in Congressional elections. This legislation, still pending, would permit auxiliary private contributions to supplement the stated amounts of federal aid.

The Panel supports this legislation in principle.

The issue of public support for independent and minor party candidacies must be recognized. Americans generally favor a two-party rather than a multiparty political system, yet few would wish to suppress strong independent candidacies. Qualification for public funding can be based upon matching a minimum level obtained in small gifts from private sources, or upon filing petitions signed by a high percentage of eligible voters, or upon payment of a sizable filing fee.

The Panel supports the rights of minority parties to participate in any public funding scheme, subject to appropriate limitations.

ASSUMING PARTIAL PRIVATE CAMPAIGN FINANCING: SOURCES AND AMOUNTS

The right to contribute financial support to a candidate or a political party is viewed as an extension of the personal or individual participation by citizens in the political process by voting or otherwise. This right can hardly be claimed by corporations, organizations, or associations which have no voting rights and which normally include individuals of opposing political views. Political gifts by corporations, as such, have long been forbidden by law in federal and in most state elections. Gifts by labor unions, as such, in federal campaigns have been illegal for a quarter of a century. Violations have occurred in both cases, but have not been vigorously prosecuted until quite recently.

There is strong reason to continue these two prohibitions, with more severe penalties and vigorous enforcement. The evasive device, used by both corporate and union leadership, of establishing parallel campaign finance units, theoretically or technically independent, with pressures applied to organization members to contribute, should also be forbidden. There is no clear distinction between such schemes and the direct use of money in campaigns by such organizations as the American Medical Association, Dairymen's Association, and scores of other special interest groups wishing to influence governmental policy by making gifts to candidates or parties. All of these sources are highly suspect, if we sincerely desire to remove the most obvious means for corruption of the democratic process.

With partial public support lessening the need for private funding, donors to campaign funds should be strictly limited to individual United States citizens of voting age. This would eliminate organizations and associations of all kinds from this process, except for the political parties. Groups or associations formed for eco-

nomic or professional reasons should not be allowed to contribute funds in federal elections. Severe penalties for violations should be provided.

Heavy borrowing by candidates or political parties can have extremely damaging long-term effects on the political process, as witness the bankrupt condition of the Democratic Party following the 1968 campaign and continuing to the present date. Large creditors are able to exert undue influence upon party policies. The fact that a victorious party can readily and quickly pay off large debts is in itself a devastating commentary on the prevailing condition of campaign financing.

Provision of partial public financing should justify legislation limiting and controlling borrowing for campaign purposes.

LIMITS ON AMOUNTS GIVEN

Strong arguments have been advanced against any limitation on amounts that may be given by any citizen. The Committee for Economic Development, The American Enterprise Institute, and many others have taken this position. There is a conceivable Constitutional issue, since any limitation on personal freedom is open to challenge. But Article I, Section 4, gives Congress explicit power to regulate "The times, places and manner of holding elections for Senators and Representatives." This would presumably govern. Although Presidential elections are not covered by this section, reasonable limits would probably be sustained if found necessary to preserve the integrity of political and governmental institutions.

There is Panel consensus that a statutory limit should be placed upon campaign gifts by any individual voter during the course of any calendar year.

This limit might well be $10,000 in any Presidential campaign, $3,000 in any Congressional campaign, with an overall maximum of $25,000 by any family in any calendar year.

One argument advanced in favor of private campaign funding is that the poor and persons of modest means can contribute their time and personal energies to political campaigns, whereas the wealthy and well-to-do are so busily occupied with their own affairs that money gifts may stand in lieu of direct involvement. But the vast discrepancy between the monetary value of donated services and the larger campaign contributions serves to undermine the validity of this argument.

Limits on total expenditures in each primary and final election campaign are a rational corollary to partial public financing—and to limits on individual donations. Such overall limits on spending could be reasonably set at the amount of public appropriations, in each case, plus an equal amount of private donations. Then public and private support would roughly balance each other.

Limits should be observed by the candidates themselves, as well as by their constituents. In other words, each candidate should be permitted to contribute only to the maximum allowed for his competitors.

This limitation would be more severe than that of the 1971 law forbidding any expenditure over $50,000 in personal funds by a Presidential candidate in a primary election. Admittedly, personal expenses for travel, lodging, food, and other ordinary needs could not be made subject to this rule.

The arguments for and against strict limits are well known. The view has long prevailed that a man should be permitted to use his own money for political purposes as he may see fit. On the other hand, persons without great wealth are severely handicapped in Congressional as well as Presidential contests, and it is certainly no part of American tradition that public office should be limited to persons of wealth. Again, public financing of campaigns at a minimum level would help to provide reasonable opportunity for outstanding citizens of all walks of life to contest for any office of public trust.

This country cannot permit its basic electoral processes to be subverted or corrupted, either in fact or in the minds of its citizens. Confidence in the integrity of the nation's governmental institutions is at low ebb; survival requires its restoration.

97

Solicitation, Form, Timing, and Conditions
of Campaign Gifts

The proliferation of campaign committees in federal elections has reached ludicrous proportions. There were nearly 5,000 of these in the 1972 campaign, most of them mere paper fronts designed to avoid gift taxation or conceal dubious transactions. It seems to be beyond doubt that this proliferation should be drastically curtailed. The right of each legally constituted political party organization to collect and disburse funds at federal, state, and local levels is unquestioned. Similarly, few would question the right of each candidate to establish a separate campaign fund under his or her control, but subject to public accountability.

Even this latter pattern has come into question, however, on the ground that the party organizations should be elevated to a new level of importance and authority, with individual candidacies completely subordinated to them. This view gains added weight in the light of the 1972 experience with the Committee to Reelect the President, separated from the Republican National Committee. Presidential candidates can usually control their national party organizations, of course, but Congressional candidates frequently do not enjoy comparable status with state or district organizations. Hence, separate campaign funds for Congressional candidates are more clearly justifiable.

But what rational justification can there be for any nonparty organization or for any noncandidate or nonpolitical grouping to collect and spend campaign funds? Freedom of expression and of association are Constitutional rights, implying that any group may announce its political views. But an implicit right of any group to spend money to influence the electoral process is highly dubious.

The solicitation, collection, and allocation of campaign money should become solely a function of political parties and/or candidates for office.

Phillip S. Hughes, Director of GAO's Office of Federal Elections,

98

testified before Congressional committees that completely centralized control over Presidential campaign solicitations may create administrative problems, particularly as to enforcement of reporting deadlines in Presidential campaigns. But if state or local units are made branches responsible to a single national organization, with some reporting autonomy, this difficulty may be satisfactorily overcome.

Mr. Hughes has pointed out that one alternative method of controlling expenditures is set forth in the Florida Election Code. In addition to prescribing limitations on both contributions and expenditures, the Florida law requires that *all* contributions, expenditures, or obligations which are "made, received, or incurred, directly or indirectly, in furtherance of the candidacy of any candidate for public office . . ." must pass through "the duly appointed campaign treasurer or deputy campaign treasurer of the candidate." As a condition precedent to qualifying as a candidate, an individual must appoint one campaign treasurer and designate a campaign depository.

Similarly, there should be one and only one campaign fund for each Presidential candidate, and for each Congressional candidate. Political party financing can be kept separate or united with these funds. Campaign financing conducted by the four Democratic and Republican Senate and House campaign committees would and should be terminated if each Congressional candidate is limited to one single fund; the national political parties could assume their functions.

The use of large sums of cash kept in suitcases, office safes, and safety deposit boxes makes effective control of campaign financing difficult. Inevitably, it also provides room for scandal and suspicion.

Every gift to a campaign fund in excess of $10 should be by check or money order, and each donor should be clearly identified. This implies prohibition of the use of "conduit" channels designed to conceal actual sources of funds. Further, gifts of securities that have appreciated in value, in order to avoid payment of capital gains taxes by the donors, are clearly unsound in terms of public policy and should be prohibited.

Disclosure of Private Contributions to Incumbents, Candidates, and Political Organizations

The concept of reporting campaign contributions appears to have almost universal acceptance in principle, and is applied more strongly than ever before in federal legislation which became effective April 7, 1972. However, administrative problems have developed, particularly as to the timing of public reports on specific dates prior to elections. These problems are in process of solution, which can be achieved through computerized reporting of all gifts, with identification of individual donors by social security numbers as well as by residential and office addresses. Prompt and complete reports could then be released for public information.

Enforcement should be strengthened by a statutory requirement that donors making campaign contributions in excess of a certain sum (perhaps $100) in any calendar year when a federal election is held would have to list and report such gifts independently to the electoral commission recommended below.

This would permit a cross-check against the reports made by those in charge of campaign funds.

Time Frames of Campaign Gifts

Collection of campaign funds long in advance of an election may lead to gross improprieties. At the Presidential level, Mr. Nixon "carried forward" from the 1968 campaign large sums of money, presumably for eventual use in his own reelection in 1972, although a portion may have been diverted to certain 1970 campaigns. It is a frequent practice for members of Congress to accumulate campaign funds long in advance of election. Such accumulations lend themselves to dubious purposes involving, for example, the personal as opposed to the campaign needs of a potential candidate.

It is easy for an incumbent to obtain financial support from special interest groups and lobbyists. Tickets to "testimonial" functions find a ready market. Incumbents enjoy a great advantage, since such sources are far less available to potential opponents. These "boosters' funds" have less justification today than they did when Mr. Nixon's fund was publicized in 1952 or when Senator Dodd was censured in this connection; Congressional pay scales have been sharply increased, along with more liberal allowances for travel, telephone expense, postage, and so on. Partial public campaign funding would eliminate any conceivable justification for this kind of "carry-forward" fund.

Accumulation of campaign funds by any candidate should be prohibited entirely except within some reasonable time frame (perhaps six months) before the primary or general election that is to be contested. Political parties, of course, would be free to collect and accumulate funds at any time for support of their ongoing operations and for use in final elections, subject to strict disclosure requirements. And provision should be made for strict accounting of candidates' campaign funds—whenever accumulated—to prevent commingling of personal with campaign expenses.

Disposition of Surplus Funds

Monies left over after payment of all campaign expenses are now seen to pose serious problems. If carried forward in private hands, for disposition without effective supervision or control, their use will be subject to suspicion even if not diverted to the personal financial advantage of the holders. Three obvious alternatives present themselves.

First, the surplus funds could be returned to the original donors, on some pro-rata basis, although this would involve material administrative difficulties. Second, all surplus funds collected for the benefit of individual candidates could be turned over to the official party organizations shortly after the primary or final election for which they were raised. Third, such funds could escheat to the government—a solution which will have more obvious merit in the event that public financial support is extended to candidacies for federal office in the first instance.

The preferred solution is for surplus funds controlled by individual candidates for federal office to be escheated to the U.S. Treasury if not transferred directly to a national or state party organization within 60 days after the election. An alternative arrangement would be to make surplus funds not transferred to party control personally taxable to the candidate if held in excess of the 60-day limit.

Administrative Mechanisms for Effective Enforcement

In the past, primary responsibility for enforcement of reporting and related matters was placed in the hands of the Clerk of the House of Representatives and the Secretary of the Senate. These officials were required to report apparent violations to the Attorney General of the United States for prosecution. Such reports were seldom made, however, and when made they were never prosecuted, prior to 1973. These arrangements were obviously deficient.

Legislation of 1971 placed primary enforcement responsibility in Presidential campaign finance upon the General Accounting Office, which established an Office of Federal Elections for this purpose. There, for the first time, a determined effort to establish and manage a thoroughly effective reporting system was launched. However, the enforcement authority of this Office was limited to taking evidence of violations to the Department of Justice, which —again—has failed to undertake vigorous prosecutions of delays or inadequacies in fund reporting. Some convictions for violation of the ban on corporate giving have been obtained, and mild penalties imposed, but only after a civil suit by a private civic organization had forced the disclosures leading to prosecution. Moreover, reports on Congressional campaign finance still go to the House and Senate officers, as before.

Testifying before Congressional committees on ways to strengthen enforcement mechanisms, Mr. Hughes commented upon two alternatives: (1) creation of an electoral commission composed of highly respected citizens serving on a part-time basis, to be appointed by the President, with a full-time director and adequate staff wielding broad authority; and (2) strengthening of the existing mechanism in the General Accounting Office, supplemented by a seven-member advisory board representative of the major political parties, the Congress, and the President.

The great weakness in past efforts to enforce the laws has been the reluctance of the Department of Justice to prosecute violations. Earlier in this report, the Panel discussed the creation of a permanent office of Special Prosecutor. With the creation of such an office, we would anticipate that more vigorous pursuit of both nonfeasance and malfeasance in the electoral process would result.

The need for strong, independent enforcement mechanisms is clear. In his testimony of November 1973, Mr. Hughes said:

> An alternative to vesting prosecuting authority in GAO would be to authorize the Attorney General to appoint a special prosecutor within the Justice Department for a specific term, subject to confirmation by both houses of Congress and removable only by Joint Resolution of the Congress for specified causes. This special prosecutor would be authorized to investigate and prosecute criminal violations arising from matters referred by the GAO and other matters arising out of the election campaigns. While an appointee of the Attorney General, he would not be removable by him or by the President. He would nonetheless have available the full resources of the Justice Department and the Federal Bureau of Investigation as well as the U.S. Attorneys. I believe this arrangement would attract a high caliber lawyer to serve for the time required to carry out such duties for each election campaign.

The Panel agrees with the recommendations that would: (1) create an electoral commission; and (2) assign prosecuting authority for cases referred by the electoral commission to the Permanent Special Prosecutor, proposed in Chapter 4.

* * *

The kinds of abuses in the financing and conduct of federal elections associated with Watergate cannot be permitted to go uncorrected. As stated earlier, an open, free, and honest electoral system is fundamental to a democratic society. An ideal system may not be achieved in one wave of reform, but the Panel's recommendations, if adopted and implemented, will provide the basis for an electoral system in which the public can have confidence. Citizens can then derive a truer sense of participation than has been possible for many decades.

The effective administration and enforcement of the proposed reforms are, obviously, critical to their success. A solid beginning has been achieved by the auspicious record of the General Accounting Office's Federal Elections Office; the proposed electoral commission should build upon that excellent base.

CHAPTER 8

Congressional Oversight of the Executive Branch

ARTICLE I, Sections 1 and 8 of the Constitution makes clear that the central function of the Congress is to legislate. Inherent in this legislative power to authorize programs, create agencies, and appropriate funds is a corollary responsibility to exercise control over the agencies of the Executive Branch. The extent and the manner of such control has been a continuing issue between the Congress and the Chief Executive.

Authority to determine the objectives of executive action implies some authority and responsibility to insure that adequate steps are taken to achieve those objectives. Accordingly, Congress must be concerned with the *how* as well as the *what* of executive action. It must consider how policies are being executed, whether they are accomplishing the desired results, and, if not, what action Congress should take. These are the activities subsumed within the responsibility of Congressional oversight.

The Oversight Function

Responsibility for exercising this oversight authority is shared by individual members of the House and Senate and by regular and special committees of the Congress, supported in their efforts by the General Accounting Office and the Congressional Research Service.

ROLE OF THE INDIVIDUAL MEMBERS OF CONGRESS

Individual Congressmen and Senators from time to time perform a monitoring role in responding to requests from constituents for assistance. This surveillance growing out of Congressional "casework" often involves criticism of agency officials for acting arbitrarily or for failing to act. This kind of monitoring is generated by criticism based on individual grievances or special interest claims, in contrast to a more comprehensive and systematic assessment designed to determine objectively whether and how well an agency is achieving its objectives.

For the most part, when performing in his committee role, the member of the House or Senate gives relatively low priority to such activity. By contrast, choosing to develop special expertise in a field of legislation within the jurisdiction of his committee and relevant to his district is much more likely to bring the member electoral support and advancement within the Congress. Exceptions can be cited, of course, when the suggestion of corruption and scandal has opened the way for a member to gain national recognition through conducting a widely publicized investigation of Executive Branch performance in a particular field of activity. But for legislative review to become continuing and meaningful, the two houses of Congress must assign high priority to this function. Only then is the average member likely to accord such activity greater attention.

ROLE OF COMMITTEES

The *legislative committee* develops special competence through the experience of its members in carrying through from idea to passage the authorizing bills for programs within its jurisdiction. Typically, its members maintain regular contact with agency program officials, making informal suggestions and checking on gains achieved and new problems encountered. The special knowledge and privileged access of these committee members qualify them to carry the formal responsibility for program evaluation. Yet they may also be more susceptible to the influence of organized interests which have a significant stake in protecting the program and insuring its continuation. Understandably, many members of the authorizing legislative committee emerge as staunch protectors and defenders of the agency, and for that reason are ill prepared to monitor its performance with healthy skepticism. At the other extreme, surveillance by the committee responsible for authorization bills can become so detailed that it approaches dictation of agency decisions to the point of undermining executive management.

Some legislative committees are handicapped in exercising surveillance because Executive Branch organization and Congressional committee jurisdiction are out of phase. Reorganization has been almost a continuing phenomenon in the Executive Branch, while occurring much less frequently in Congress. The result for the House and Senate is to diffuse responsibility for legislative oversight of particular departments or agencies among several committees, reducing the effectiveness of investigations and at times permitting the Executive Branch to play one committee off against another.

The Panel believes that whenever major executive reorganization is proposed, either by statute or reorganization plan, the Government Operations Committee in each chamber should review and report how committee responsibility for oversight would be affected.

The *Government Operations Committees* have clear statutory authority to investigate any government activity, even when such exploration involves cutting across agency jurisdiction, program, and policy lines.[1] By contrast, legislative committees are expected to stay within their defined jurisdictions. Despite their broad

[1] Section 136, Legislative Reorganization Act of 1946 (60 Stat. 832). Some discussions of Congressional oversight draw a distinction among legislative oversight, investigative oversight, and fiscal review. The first refers to review by the authorizing committee of programs within its jurisdiction, usually leading to new legislation. Investigative oversight is considered more exploratory, broader in scope, often without a specific legislative objective, and more likely to have been launched in an effort to uncover wrongdoing. The role of the Government Operations Committees fits this type of oversight. Fiscal review is concerned with monitoring budget execution and normally involves the Appropriation Committees. See Walter Oleszek, "Congressional Oversight: Methods and Reform Proposals," U.S. House of Representatives, 93rd Congress, 1st Session, Select Committee on Committees, *Panel Discussions before the Select Committee on Committee Organization in the House*, vol. 2, part 3 (Invited Working Papers), pp. 710–723.

investigative authority, however, the Government Operations Committees have been sensitive to possible charges of having invaded the jurisdiction of legislative committees. At times, House legislative committees have challenged the Government Operations Committee on jurisdictional grounds, adding to the confusion surrounding the exercise of oversight responsibilities. Clearly the Government Operations Committees occupy a central role in oversight; nevertheless, overlapping and diffused responsibility limits their effectiveness.

The *Appropriations Committees* probably have engaged in more continuous surveillance of executive agencies over the years than have any other committees of the Congress. Often such monitoring is in the form of requiring that certain proposed decisions or expenditures be cleared with the appropriations subcommittee in advance. The Appropriations Committees use a variety of both formal and informal devices to guide agency actions, including: (1) specifying the purposes for which money may be spent; (2) prohibiting the spending of money for specified purposes; and (3) modifying the level of funding available for particular programs, even to the point of eliminating a program by cutting off all funds. These instructions take many forms. They may be expressed in appropriation act language, in reports from the appropriations and conference committees to the two houses, and in the hearings themselves.

Even at their best, however, the two Appropriations Committees (and their many subcommittees) are not adequately equipped by formal charge, by organizational structure and style of operation, by time, or by staff to undertake the broad review of Executive Branch performance and policy which is most needed.

Ad hoc Committees of Congress also have made important contributions in the conduct of legislative oversight. At best, however, the effectiveness of joint, select, and special committees is spotty. Their principal weakness in exercising surveillance is lack of continuity, since these committees have a limited life span. There will always be the unusual situation in which an ad hoc approach can achieve notable success, but for the regular performance of program appraisal the ad hoc committee can be no more than a useful supplement.

In 1970 Congress significantly strengthened its oversight capability by specifically directing the *Comptroller General* "to review and analyze the results of government programs and activities carried on under existing law, including the making of cost budget studies" as a means of assisting Senate and House committees in their efforts to evaluate executive agency programs.[2] The General Accounting Office has responded to these new responsibilities by expanding and enriching its professional staff, drawing about a quarter of its 3,250 professionals from disciplines other than accounting, including computer science, economics, engineering, industrial management, social sciences, statistics, and systems analysis.[3] The Congress and its committees now have access to greater staff resources and more competent staff than ever before to support the oversight function.

Summarizing this brief review, the modern Congress does not yet devote enough attention to improving its ability to oversee the Executive Branch. Congressional committee efforts to review program performance are infrequent, and rarely comprehensive. They often become a search for evidence to justify curtailing a program, in contrast to assessing it objectively. Staff members of the House and Senate (including the General Accounting Office and the Congressional Research Service) pursue oversight issues when asked by members to do so, and these requests are growing.

Strengthening Congressional Oversight

The Panel believes that the central objective of legislative oversight should be to hold Executive Branch officials accountable for the honest and ethical performance of their responsibilities, and for attaining program objectives defined in authorizing and appro-

[2] Legislative Reorganization Act of 1970, 84 Stat., p. 1168.
[3] See Elmer B. Staats, "General Accounting Office Support of Committee Oversight," in House Select Committee on Committees, *Panel Discussion on Committee Organization in the House*, vol. 2, part 3, pp. 692–700.

priating legislation. The principal mechanism to give this accountability meaning is the instrument of performance review and evaluation. But evaluating program performance is extremely difficult, and both the Executive Branch and the Congress are still at an elementary stage in developing sound standards of measurement and effective ways to audit performance.

Watergate events have demonstrated the crucial importance of continuing and systematic legislative overview of the Executive Branch, including the White House and the Executive Office of the President. Such continuing evaluation could have made members of Congress aware much earlier of the significance and dangers of many of the developments discussed elsewhere in this report.

A continuing feedback of information about Executive Branch performance can provide Congress with a better opportunity to question, consider, debate, and, perhaps, modify—by influence or by legislation—developments affecting the evolution and the functioning of our governmental institutions.

Against this background, therefore, the Panel recommends that the Congress give major attention to strengthening its capacity to perform the oversight function, with particular emphasis on evaluating the performance of executive agencies.

CLARIFYING OF RESPONSIBILITIES

The Panel has reviewed alternative proposals to fix more clearly the responsibilities for oversight of the Executive Branch within the Congress.

First, it urges that the House of Representatives and the Senate each develop, discuss, and adopt a clear statement of the assignment of responsibilities for administrative oversight.

The Panel has considered the proposal advanced by some for vesting in a single committee in each chamber, or in a joint committee, full authority for evaluation of Executive Branch programs. Performance of oversight, however, is too closely tied to authorization of programs and appropriation of funds to be separated completely from the responsibility of legislative and appropriations committees. Moreover, the Panel does not believe that one committee in each house or a single joint committee can handle the potential workload required to discharge this responsibility adequately. In addition, the Panel concludes that the central importance of continuing feedback from performance evaluation into the legislative process militates against a joint committee on oversight. The House and Senate must each develop its own oversight capability.

Furthermore, the Panel finds merit in the conception that the oversight function involves three related but distinct dimensions: legislative review, fiscal review, and investigation. To clarify where responsibilities should lie, the legislative committees should be assigned the job of monitoring agency performance to measure progress toward attainment of program objectives set forth in the authorizing legislation. This is the dimension of legislative review.

To insure full attention to agency program performance, the Panel urges serious consideration of the proposal made by the House Select Committee on Committees that each legislative committee establish a standing subcommittee on oversight.

These subcommittees would carry out surveillance of programs authorized by legislation within the jurisdiction of their parent committees. A standing subcommittee on oversight is likely to provide somewhat more objectivity in its approach to surveillance than would a subcommittee with substantive responsibility for managing authorization bills.

Primary responsibility for the second dimension—fiscal review —should continue to rest with the Appropriations Committee in each house. The budget reform proposals which have emerged after lengthy study in both House and Senate will, if adopted, require the Congress to undertake a more comprehensive and integrated process of fiscal planning and budget review than now exists. To implement the kinds of changes proposed will in turn necessitate more systematic, regular, and continuing analysis of agency performance as a means of evaluating budget requests. Congress can strengthen its capability for assessing agency performance by relating program evaluation carried on as part of legislative review with the fiscal appraisal regularly performed during the budget-appropriation process.

Developing effective coordination, however, between legislative committees and appropriations subcommittees in their view of agency programs will be most difficult. The built-in differences in objectives and perspectives between the two types of committees pose serious problems. Yet unless there is a close working relationship, pooling of data and findings, and open discussion and sharing of insights among members of these committees, effective performance evaluation cannot be achieved within the Congress.[4] The Panel believes that the approach proposed for strengthening the oversight capability will also reinforce efforts to develop a more integrated process of budgetary planning.

The Panel endorses the proposal of the House Select Committee on Committees to vest the Government Operations Committee with the central role of coordinating and monitoring the performance of the oversight function in the House. The responsibility of these two Committees for conducting broad investigations of the Executive Branch should be confirmed. The Panel further urges that the resolutions setting forth the jurisdiction of these two Committees explicitly recognize the assignment of responsibility to monitor the organization of the Executive Office of the President and the White House.

Congress has been understandably reluctant to prescribe how the President should organize his own immediate office or to limit his choice of that arrangement which best fits his approach to executive management. Nevertheless, in the judgment of the Panel, it is entirely reasonable that the President be asked to

---

[4] Allen Schick, Congressional Research Service specialist in budgeting, observed in a paper prepared for the House Select Committee on Committees: "For program evaluation to work requires that budgetary incentives be structured in its behalf. No matter how much they want good programs, members of authorizing committees now find it easier to seek additional funding for new programs than to review the achievements of programs already under way. As long as the Appropriations Committees are the watchdogs, the authorizing committees will find it hard to curtail their support for programs under their jurisdiction. What this suggests is that it will take some fundamental changes in the relations between the Appropriations and the legislative committees before program evaluation can bloom on Capitol Hill." House Select Committee on Committees, *Panel Discussions on Committee Organization in the House,* vol. 2, part 3, pp. 627–628.

provide the two Government Operations Committees explicit information about the organization of the immediate Office of the President and the assignment and delegation of responsibility and authority within that structure. Should the proposed pattern of allocating power and responsibility raise questions among members of the Committees concerning accountability and lines of authority, these issues should be considered openly through hearings and Congressional discussion, seeking to exert influence on how the President proceeds, but stopping short of legislation. The Panel favors this approach over the alternative of specifying by statute the organization and delegation of authority to officials and units in the Executive Office of the President and the White House staff.

Further, the relevant appropriations subcommittees should be provided full information about staffing in the White House Office of the President. Staffing patterns and the financing of these offices should be discussed openly and supported by the same kind of justifications required of other Executive Branch agencies.

DEVELOPING A CONTINUING SYSTEM

Through the initiative of its Government Operations Committee, each house would consider and adopt a defined agenda at the beginning of each Congress, outlining program priorities to be pursued in performance evaluation activities, committee by committee, during the two-year life of the Congress. The Government Operations Committees would oversee coordination and scheduling of these performance review studies.

In addition, the Government Operations Committees should have authority to see that relevant findings from their own investigations, from legislative committee reviews, and from appropriations subcommittees' fiscal reviews are fed into the legislative process and the appropriations process at the right times.

Once this systematic approach to program review has taken hold, there may be enough business generated to warrant regular scheduling of oversight issues on the floor of the House and Senate once or twice a month.

Assignment of program evaluation studies to the General Accounting Office and requests to the Congressional Research Service to undertake relevant background explorations could also be coordinated through the Government Operations Committees. In short, these two committees can be given authority to manage, guide, and coordinate performance of oversight tasks, while also serving as an information clearing house for materials on how the Executive Branch is performing.

The Panel believes that building such a system into the regular day-to-day operations of the House and Senate will give higher priority to the oversight function. As the system gains acceptance through support from party leadership and committee and subcommittee chairmen, the incentive for individual members to participate actively is likely to grow. The impact of such continuing performance evaluation upon the Executive Branch can also bring important by-products. Congressional demands for regular reporting of program information, plus the anticipation of periodic Congressional committee review, should stimulate agencies to strengthen their own program management and improve internal systems of reporting and performance assessment.

The Panel believes that members of Congress and committee staffs can make more fruitful use of the program audit reports prepared by the General Accounting Office. For example, careful review of these reports by the Government Operations Committees and their staffs should provide insights helpful in guiding coordination of legislative and fiscal reviews. The Congressional Research Service also can provide research support to help sustain significant expansion of Congressional review of Executive Branch programs.

Finally, the Panel favors continuing attention by Congress to further strengthening of its research and analytic capability, since the proposed approach to oversight will severely tax the supporting resources presently available to Congress. Proposals to augment staff and to develop information management resources deserve serious consideration, because they would help to facilitate more systematic surveillance of the Executive Branch.

IMPROVING ACCESS TO INFORMATION

Effective legislative review is also dependent upon the free flow of information within and throughout the federal government. Yet the Watergate hearings offer substantial testimony to suggest that barriers and distortions in that flow have reached an unacceptable level. The claim of executive privilege appears to have been employed to frustrate efforts to seek out wrongdoing. On the other side of the coin, elaborate public relations programs in some agencies blanket Congressional committees with more—and often irrelevant—"information" than they can possibly handle, befogging issues and distorting facts in the process.

While rejecting proposals for major Constitutional reform in the direction of the parliamentary system (see Chapter 1), the Panel did devote serious attention to the feasibility of a regularly scheduled question period as a mechanism for improving information exchange and strengthening Congressional oversight. Essentially, such a procedure would involve setting a time when President, Cabinet officers, and/or other agency heads would appear before the Congress (in a joint session or in one or the other chamber) to respond orally to questions submitted in advance and then to answer follow-up questions from the floor. Requiring the President to appear would necessitate a Constitutional amendment. Cabinet officers and agency executives now appear before legislative committees from time to time, with generally productive results. In the Panel's judgment, such a formal question period would not serve the essential information function. At best it would become a diverting "side show," and at worst it could produce mischievous results, making or breaking political reputations, often for irrelevant reasons.

The Panel, therefore, counsels against trying to adapt the question period to the use of the United States Congress.

Another important aspect of information access is the improved management of information. The heart of the proposed process of Congressional oversight is evaluation of the integrity and effectiveness of administrative performance, and the capacity to appraise program performance in turn requires regular and full reporting. As evaluation becomes more systematic and comprehensive, the flow of information will increase exponentially. The Congress must be prepared to handle this information efficiently as an essential step to carrying out administrative oversight honestly and constructively. The growing capability of the General Accounting Office and the Congressional Research Service in information management is encouraging. But Congress must be alert to find ways to improve information management; this will require continuing attention.

Congress must also work to overcome the tendency of some administrators to protect internal operations from outside pressures and influences behind the cloak of security. In addition, the Congress may soon reach the point where it needs a specialized legal service (or office of Congressional counsel) to press for information denied by the Executive Branch. The task of employing legal processes to gain access to information can of course be handled by counsel to a standing committee, but if Congress finds that it needs such a specialized legal service to handle a large volume of legal troubleshooting for its standing committees, such an office should be considered.

* * *

In its long history the Congress has a mixed record in performance of the oversight function. This is fully understandable, for it is an exceedingly difficult responsibility to perform well in an institution which by design is intended to represent many diverse interests. Development of an institutionalized system of Congressional oversight, as here proposed, is a gradual process. It will not spring into being full-blown. Neither will it be achieved without a clear and firm commitment to move in this direction, followed by careful nurturing until the benefits of a more systematic approach to oversight become apparent.

If the budget reform proposals now pending are adopted, the Congress will assume significant new responsibility to synthesize its painstaking analysis of budget details with the formulation of comprehensive national budgetary policy. The thrust of this Panel's recommendations regarding oversight calls for the same kind of effort—to develop a system of monitoring the Executive Branch, harnessing detailed evaluation of program performance to the attainment of broad policy goals. This is a difficult assignment, but the potential rewards for the country, and for the Congress and the Executive, are great.

CHAPTER 9

The Presidency and the Judiciary

IN the events following the Watergate burglary, the Judicial Branch of the federal government was called upon to decide two kinds of disputes: (1) criminal prosecutions by the Executive Branch; and (2) controversies involving the extent of particular privileges and immunities of the Chief Executive. The common feature of the cases, not surprising in view of the circumstances, was the exceptional nature of what the courts did, or almost had to do.

Criminal Prosecutions by the Executive Branch

Two disputes between the Executive Branch and individuals, criminal prosecutions in the federal courts, helped expose the larger Watergate affair. In each case the courts behaved in exceptional ways.

Had the trial of the Watergate burglars been an ordinary prosecution of unsuccessful and nonviolent thieves without criminal records, what the judge did would surely have been considered to stem from an inquisitorial bias. The judge's persistence was, of course, based on a manifest concern that the facts were not being brought out in the courtroom. After conviction, he imposed sentences which, in an ordinary burglary, would have been considered extremely harsh. He imposed them conditionally, again out of a manifest desire to expand the inquiry by pressing the defendants into talking.

In the other prosecution, that of Daniel Ellsberg and Donald Russo, the trial judge dismissed the indictment on the basis of the general impropriety of the government's behavior, apparently without regard to whether any specific impropriety had affected the case as actually put before the jury. Had this been an ordinary prosecution, the judge's action would surely have been considered a significant extension of the precedents for dismissal because of illegally obtained evidence or because of flagrant misconduct by the prosecution.

In those two prosecutions, the courts pushed to the very edge of normally proper judicial conduct in response to even more extreme behavior by another branch of the government and in order to secure a civic objective of overriding importance. One may reasonably suspect that the two courts ultimately saw the real matters before them not as burglaries or unauthorized disclosures, but as attempts by the Executive Branch to subvert, respectively, the electoral process and the judicial process.

The Powers of the Chief Executive

The main disputes about the limits of executive power concerned the Presidential tapes. Those litigations came a year after the criminal prosecutions discussed above and well after much of the Watergate case was broken. They involved the familiar effort by courts to secure for their own use the best possible evidence. Despite their air of ultimate confrontation, the later cases proceeded along more traditional lines. Their real tension was in two questions: (1) whether the courts could, and should, decide the limits of the President's ability to withhold information; and (2) whether the orders of the courts would be obeyed.

Despite the absolutist ring of some of the arguments, the main question presented, and decided, was not whether, under the doctrine of separation of powers, one branch had plenary power to declare the extent of its own power. Indeed, on that simplistic level, the entire outcome would seem to turn on whether one posed the question in terms of the President's deciding what he could withhold or in terms of the court's deciding what it could require.

Instead, what was decided was that: (1) an officer of the Executive Branch, including the highest officer, had to obey the court's orders for the production of evidence in the Executive's possession which was needed by the court (or its instrument, the grand jury) to determine whether particular individuals had violated particular laws; and (2) the court was the body to balance competing interests, such as the President's need for frank advice without fear of later compromise or revelation, and the courts' need for the best evidence. Before drawing broad conclusions about that decision as a precedent, it is worth noting that the decision was taken in a specific criminal case in which: the evidence promised to be highly relevant; privileges had already been waived for purposes of testimony; the evidence was of the President's creation and related to his own behavior and that of his closest associates; and the President forced the court's orders to run to him by publicly taking personal custody of the tapes. That was an extraordinary cumulation of circumstances weakening the President's claims to immunity or privilege.

At this writing, the courts have not decided the parallel question of whether the courts could, and should, decide what the President could withhold from a Congressional inquiry. Initially the court avoided the question on jurisdictional grounds, but Congress corrected the defect and committee subpoenas are now before the courts for enforcement.[1] Without in any way prejudging this litigation, the Panel notes that for a court to order a Presidential compliance with Congressional subpoenas raises hard questions. The need for the information may be less than in a

---

[1] On February 8, 1974, District Judge Gerhard Gesell ruled against the Select Committee's petition. The District Court decision is at this writing on appeal to the U.S. Court of Appeals for the District of Columbia.

criminal trial; alternative information may be available upon which to base legislation; Congress has other ways than court process of dealing with an uncooperative Executive; and the danger of fishing expeditions and other abuses may be great.

On the narrow question of who decides what evidence the Executive must turn over to the court in pending criminal cases which are properly before the court and to which the evidence is highly relevant, it seems entirely correct to the Panel that the final decision as to the extent of this particular executive power must rest with the courts.

How to administer that arrangement, in such delicate areas as executive privilege and national security, is discussed elsewhere in this report, and the Panel does not here mean to suggest that the interests of Presidential confidentiality should not be given proper weight.

It is worth noting that, while the question resolved in the tapes cases related to only one particular assertion of executive power, it is not the only question of executive power which the courts must, from time to time, and quite properly, resolve. What defines those situations is not their content but whether they arise in a justiciable case—an instance of real dispute between real people over whether actual behavior violated existing law.

The other uncertainty in the litigations over the tapes—whether the direct orders of the courts would be obeyed—concerns something which might have happened, and at one time appeared likely to happen, but which, at this writing, fortunately did not. Raw powers abound in our governmental system which, if pushed to the extreme without regard for legitimacy or for the distinction between extralegal power and authority, could mock or destroy our system. The Congress could refuse to appropriate funds for the White House salaries or food bill. The courts could issue advisory opinions on hypothetical cases and command general compliance. The President could refuse to enforce laws distasteful to him, defy court orders, or rely, not on the Constitution, but on the obedience of the armed forces.

But the equivalent of those things did not happen in the tapes case. What would have followed if the President had defied a court order directed to him we do not know. But specific court decrees directed at an individual have great moral force in our system, at least as concerns that individual's duty to comply. Even if one considers such legal questions to resolve, in the extreme, into political questions, a defiant President would need vast reserves of public trust, as well as less admirable attributes, in order to defy and survive.

Judicial Self-Restraint

The strain on our institutions symbolized by Watergate has, of course, been felt by the judiciary. Much of what the courts did was, apart from the dramatic context, routine, and fortunately so, for when it was not, the courts were able to fulfill their larger role in safeguarding the integrity of other institutions only by pressing their own powers to the utmost. Just as the courts have occasionally stepped beyond their classical jurisdiction to make law which a more appropriate branch was considered inherently incapable of making, so here the courts stepped beyond their customary patterns and restraints in order to make sure that the law was obeyed by those constitutionally entrusted with executing it, especially where the broken laws concerned the balance of powers among the agencies of government or between the government and the people.

The courts may have to do so again, for what they did here was serve as a surrogate for self-restraint. Modern American culture is quite capable of producing leaders who are intensely pragmatic, oriented to success, and oblivious or scornful of institutional tradition. And that is when the courts may again have to transcend their own traditional proprieties.

These are not, however, the sort of heroics to be expected of, or entrusted to, the judiciary on a continuous basis.

The actions of the courts in Watergate were proper, even vital, but they should not be taken as showing a need for greater judicialization of governmental procedures, for more court in-

volvement in the primary business of the other two branches, for a transfer of traditional functions, or for casual recourse to the courts to reconsider decisions of the other two branches merely because they are thought unwise. Certainly no court should be lured into the heady and unmanageable activity of advising on law without facts or without the discipline of having actual controversies before it.

The judicial process, the adversary system of reaching decisions, legal reasoning, and case law as a system of social control all have serious limitations as ways of governing a complex, modern society, and nothing in the Watergate affair teaches otherwise.

The crucial role of the judiciary in putting Watergate to rights is, in the view of the Panel, evidence mainly that its role in our government is soundly conceived, not that that role should be significantly expanded or that the judicial process be taken as a model for other areas of government.

Certainly a vigilant judiciary is no adequate substitute for honesty, competence, restraint, and civility elsewhere in government. It is more likely that only in the context of those qualities in government, and of the belief in them by the people, can the courts effectively defend such fundamental qualities of our society, the government, and the people from particular attack.

EPILOGUE

Ethics and
Public Office

MOST of this report has concerned, directly or indirectly, the subject of ethics in the public service. So did most of the hearings before the Senate Select Committee. The investigative power of the Congress may well be a more effective instrument than the criminal procedures of the courts in exposing, and thereby protecting the public from, unethical behavior on the part of its officials.

The only thing that could and can avail the body politic *in extremis* is the charter of the Select Committee, committing it to the investigation of unethical—not just illegal—conduct in the 1972 campaign. The unethical is not necessarily—not even often—the illegal, as Congress attested in separating the two. It was the unethical, not the illegal, activities in 1972 that did this country down. . . .[1]

Many of the actions associated with Watergate, the burglary of offices, the forgery of a letter, the laundering of money through Mexico, and so on, were clearly criminal. But in their relation to the national interest each by itself was less than crucial. What was important were the attitudes of mind, the modes of conspiring, and the narrow goals of those behind them. Many of these kinds of matters lie beyond the range of criminal law.

Public officials are of course bound by the same criminal laws as apply to other citizens. But their obligations to the public as a whole entail an additional and more rigorous set of standards and constraints associated with the concept of public trust. Many prac-

[1] Milton Meyer, "From Deliquescence to Survival—Watergate and Beyond," *Center Report,* Center for the Study of Democratic Institutions (February 1974), p. 27.

123

tices which are permissible, even normal, in the private sector are, or should be, forbidden in government: acceptance of certain kinds of gifts, discussion of appointments under certain circumstances, promise or threat of governmental action under some circumstances, carrying and secreting of large amounts of cash, withholding of information to which the public should be alerted, and, conversely, leaking or other disclosure of other kinds of information which should be private.

One of the characteristics of many of those implicated in Watergate was their perception of the roles and responsibilities of government, a perception which was at best simplistic, and at worst venal and dangerous. A democratic government is not a family business, dominated by its patriarch; nor is it a military battalion, or a political campaign headquarters. It is a producing organization which belongs to its members, and it is the only such organization whose members include *all* the citizens within its jurisdiction. Those who work for and are paid by the government are ultimately servants of the whole citizenry, which owns and supports the government.

Complementary to the ingenuousness of the appreciation of the sense of the word "public" in these recent developments was the apparent lack of understanding of "service." In a society in which sovereignty presumably rests in the people, it is indispensable that its officials be regarded and regard themselves as servants, not masters, of the people. They must have and exercise powers, but their powers are delegated, usually for temporary periods.

A ten-point Code of Ethics for Government Service was adopted by Congress in a concurrent resolution in 1958, and its provisions were subsequently incorporated in the *Federal Personnel Manual*. In May 1961 President Kennedy issued Executive Order 10939 as a Guide on Ethical Standards to Government Officials; it was specifically pointed to those occupying positions of highest responsibility. Many federal agencies have issued their own minimal standards of conduct, applying to the specific business of the individual agencies, and most of them have some machinery for guidance and enforcement through legal counsel or inspectorates. Yet Watergate happened.

Ethics and Public Office

The Panel has considered a number of possible steps the government might take to strengthen ethical standards, particularly of noncareer officials, and suggests them for the consideration of the Select Committee. They include:

- *Improving and making more sophisticated the codes and guidances of ethics in public service;*
- *Incorporating in the oath of office, sworn to by all new officers and employees, the Code of Ethics, and requiring each at the time to read it and certify in writing that he or she has done so;*
- *Requiring that new political appointees attend briefing sessions on the ethics of public service; such briefings would cover ethical conduct, accountability, the nature of checks and balances in government, the importance of responsiveness to the public, and the relationships between career and noncareer services;*
- *Creating a Federal Service Ethics Board, comparable to similar boards that have been established in some state and local governments, to set forth general guidelines for all employees and to investigate particularly important and difficult ethical questions that are brought before it;*
- *Providing a governmentwide ombudsman or one in each major agency to consider complaints of ethical violations in the federal service.*

The effectiveness of codes and mechanisms for their enforcement depends first upon continuing scrutiny of the decisions and actions of public officials: by their fellows in administration, by the other branches of government, by their professional associates in and out of government, by the media, and by the general public. Such scrutiny in turn hinges on openness and accountability. For those who might be tempted to unethical behavior for want of understanding or conscience, the threat of future revelation and scrutiny can be a considerable deterrent.

* * *

But there is no "fail-safe" mechanism whereby appropriate ethics of public officers and the public interest may be assured, and whereby the ethics of public employees may be enforced. Ultimately, the assurance of high standards of ethical behavior depends upon the people who aspire to and gain public office, and more particularly upon the system of values they have internalized. The Panel reiterates its urging, in the Introduction to this report, that the educational institutions around the nation, especially those professional schools which provide significant numbers of public officials, focus more attention on public service ethics. A guiding rule of such instruction and of subsequent official decisions should be that propounded many years ago by Thomas Jefferson:

Whenever you are to do a thing, though it can never be known but to yourself, ask yourself how you would act were all the world looking at you, and act accordingly.[2]

[2] In a letter to Peter Carr from Paris, France, August 19, 1785.

SAM J. ERVIN, JR., N.C., CHAIRMAN
HOWARD H. BAKER, JR., TENN. VICE CHAIRMAN
HERMAN E. TALMADGE, GA.   EDWARD J. GURNEY, FLA.
DANIEL K. INOUYE, HAWAII   LOWELL P. WEICKER, JR., CONN.
JOSEPH M. MONTOYA, N. MEX.

SAMUEL DASH
CHIEF COUNSEL AND STAFF DIRECTOR
FRED D. THOMPSON
MINORITY COUNSEL
RUFUS L. EDMISTEN
DEPUTY COUNSEL

United States Senate

SELECT COMMITTEE ON
PRESIDENTIAL CAMPAIGN ACTIVITIES
(PURSUANT TO S. RES. 60, 93D CONGRESS)

WASHINGTON, D.C. 20510

Rec'd 10/9/73

October 5, 1973

Mr. Roy Crawley
National Academy of Public
    Administration
1225 Connecticut Avenue, N. W.
Washington, D. C. 20036

Dear Mr. Crawley:

As you know, the Senate Select Committee on Presidential Campaign
Activities has been working for several months to develop facts
relating to the June 17, 1972, burglary of the Democratic National
Committee headquarters and subsequent events pertaining thereto, to
the question of "dirty tricks" in presidential campaigning, and to
campaign financing. The public hearings will soon come to an end,
at which time the Committee and its staff must turn its attention
to writing its report.

The report will be the culmination of our efforts, the most
important single matter facing the Committee. My purpose in
writing is to enlist the support of the National Academy of Public
Administration in helping the Committee identify the major
implications of the facts developed by the hearings. It would
be extremely helpful to us if you could let us have the judgment
of your organization on the following matters: (1) what are the
key issues or problems in our governmental system that have been
disclosed by the Committee's work? and (2) what are the options
or alternatives that might feasibly be open for serious contemplation by the Committee and its staff in writing its recommendations?
With respect to the latter question, what are the advantages and
disadvantages of each alternative?

I understand that you have discussed the foregoing with Arthur
Miller, who tells me of your willingness to proceed. The
Committee deeply appreciates this. The judgments of the members
of the Academy will help us focus on the major issues, particularly
since they will come from a nonpartisan group whose interest lies
in improvement of our government. You may be sure that the
Academy's product will be given the most careful consideration
by the Committee and its staff.

I join with Professor Miller in expressing regret that we did not get in touch with you sooner. But I am hopeful that you will be able to send us your results by not later than December 15, 1973. The vast governmental experience of your membership should, of course, be of great help in shortening the time required for such an important scholarly undertaking as this.

I should appreciate your confirming this arrangement by letter to me. I also think it desirable that you keep Arthur Miller abreast of your progress.

With all kind wishes, I am

Sincerely,

Sam J. Ervin, Jr.
Chairman

Biographical Summaries

Panel Members

FREDERICK C. MOSHER, Chairman of the Panel, is Doherty Professor of Government and Foreign Affairs at the University of Virginia. He is the author of *Democracy and the Public Service* (1968) and numerous other books, essays, and monographs on public administration. He has taught at the University of California, Berkeley, and Syracuse University, and has served in administrative capacities in the TVA, the Army Air Forces, and the Department of State.

ALAN K. CAMPBELL is Dean of the Maxwell School of Citizenship and Public Affairs at Syracuse University. A specialist in metropolitan government and urban affairs, in addition to his academic contributions to this field, he has been actively engaged in state and local government affairs, particularly in New York.

FREDERIC N. CLEAVELAND is Provost of Duke University and Chairman of the Board of Trustees of the National Academy of Public Administration. He served as Chairman of the Department of Political Science of the University of North Carolina from 1958 to 1970. His areas of special interest include the organization of Congress and the legislative process.

THOMAS W. FLETCHER is President of the National Training and Development Service for State and Local Government. Before assuming this position he served as City Manager, San Jose, California, and as Deputy Mayor, Washington, D.C.

During the early 1960s he served as City Manager of San Diego, California, and later became president of a food franchise corporation.

BERNARD L. GLADIEUX is a director of Knight, Gladieux, and Smith, a management consulting firm in New York City. Previously he served as an officer of the Ford Foundation and of Booz, Allen, and Hamilton. From the later 1930s to 1950 he served in several federal government executive positions in the Bureau of the Budget, the War Production Board, and the Department of Commerce.

ROGER W. JONES has served in a variety of positions in the federal government over a period of 40 years. Twenty of those years were spent in the Bureau of the Budget. He also served as Chairman of the Civil Service Commission and Deputy Undersecretary of State for Administration. He is currently a member of the board of the National Civil Service League.

HARVEY C. MANSFIELD, Sr., currently Professor Emeritus of Public Law and Government at Columbia University, has also taught at Yale University and at Ohio State University. During World War II he was in the Office of Price Administration and has served as consultant to a number of government agencies since then.

JOHN D. MILLETT, Vice President and Director of the Management Division of the Academy for Educational Development, has also served as Chancellor of the Ohio Board of Regents and as President of Miami University (Ohio).

JAMES M. MITCHELL, Director of the Advanced Study Program of the Brookings Institution, has also served as a member of the Civil Service Commission, as Deputy Assistant Secretary of Defense, and as Associate Director of the National Science Foundation.

HAROLD SEIDMAN is Professor of Political Science at the University of Connecticut. He was with the Bureau of the Budget for 25 years, serving as assistant director of management and organization during the last four of them. He is the author of *Politics, Position and Power,* written while he was Scholar in Residence at the National Academy of Public Administration.

ROBERT F. STEADMAN is a consultant to the Committee for Economic Development, where he was director of the Committee for Improvement of Management in Government for nine years. He headed the Office of Economic Adjustment in the Department of Defense and was controller of the State of Michigan. He was an executive with the American Management Association from 1956 to 1961.

RICHARD E. STEWART is Senior Vice President of Chubb & Son, Inc. He was first assistant counsel to the Governor of New York, then he became Superintendent of Insurance for the state.

Staff

BERTRAND M. HARDING, Staff Director of the study, is recently retired from the federal service. He has been associated with a number of different agencies, including the Federal Aviation Administration, the Atomic Energy Commission, and the Veterans Administration. He also served as the Deputy Commissioner of the Internal Revenue Service and the Acting Director of the Office of Economic Opportunity.

RICHARD L. CHAPMAN is a Senior Research Associate of the National Academy of Public Administration. He has previously held administrative positions with the Office of the Secretary of Defense, U.S. Bureau of the Budget, and the Public Health Service. He has also served as Staff Director for a member of Congress and Chief Consultant to the House Government Operations Subcommittee on Research and Technical Programs.

ERASMUS H. KLOMAN is a Senior Research Associate of the National Academy of Public Administration. He has served on the corporate staffs of several major corporations, with duties mainly in the area of public affairs and government relations. Prior to these assignments he served as Assistant to the Director of the Foreign Policy Research Institute at the University of Pennsylvania and as an officer in the Department of State.

INDEX

Index

McCormack Act of 1951, 33
Management and Budget, Office
of, 10, 37–38, 39, 41–43, 73
Meyer, Milton, 123
Milk fund case, 49
Mitchell, John, 56
Monocracy, 11

National security, 5, 80–81
National Security Council, 10
Nixon, Richard M., 5, 7–8, 28,
30, 31, 37, 39, 40, 42, 100,
101, 118
actions and proposals during administration, 9–11
Nixon v. Sirica and Cox, 34

Ombudsman, 125

Pendleton Act, 71
Permanent Special Prosecutor, Office of, 59, 103
Political appointees, 64–70
Politicalization of government offices:
in career service, 5
in Department of Justice, 55–62
in other government offices,
63–70
use of government powers for
political purposes, 48–51

Power:
abuse of government, 48–50
drive for, 7
President's Committee on Administrative Management, 9, 29,
38–39
President of the United States, the,
17, 24, 26
assistants to the, 8–10, 29–41
and the Attorney General, 57–
58
campaigns as incumbent by, 21,
37
and the Executive Branch, 44–
52
and the Executive Office, 27–43
term of, 22–24
Presidential pardon, 61
Presidential powers, 7, 28–32, 118–
121
delegation of, 32–34
Presidential term, 22–24
Presidential transition, 9

Reorganization Plan Number 1
(1973), 37
Reorganization Plan Number 2
(1970), 39
Revenue sharing, 9
Roosevelt, Franklin D., 29
Russo, Donald, 118

Scandals, government, 5